JUST GREEN ELECTRICITY

Helping Citizens Understand
a World without Fossil Fuels

JUST GREEN ELECTRICITY

RONALD STEIN / TODD ROYAL

ARCHWAY
PUBLISHING

Archway Publishing books may be ordered through booksellers or by contacting:

Archway Publishing
1663 Liberty Drive
Bloomington, IN 47403
www.archwaypublishing.com
1 (888) 242-5904

ISBN: 978-1-4808-9069-5 (sc)
ISBN: 978-1-4808-9070-1 (hc)
ISBN: 978-1-4808-9068-8 (e)

Library of Congress Control Number: 2020911050

Print information available on the last page.

Archway Publishing rev. date: 06/12/2020

CONTENTS

PREFACE

At a rapid pace more and more countries and governments are moving their energy policies toward ridding the world of fossil fuels. The reason is to electrify societies using only intermittent electricity from wind turbines and solar panels. **Just GREEN Electricity** – *Helping Citizens Understand a World without Fossil Fuels* brings simplicity and clarity to complex issues and educates, you the reader to be more energy literate.

Energy is more than just intermittent electricity from wind and solar. Energy and electricity include thousands of products made from petroleum derivatives that "makes things and moves products," and the various fuels required by every transportation infrastructure for prosperous societies. Energy literacy will enhance one's comprehension that the cost of energy affects everything: from the food we eat, the clothes we wear, our daily transportation, how we communicate, where we live, whether or not we have access to healthcare, and the leisurely living made possible by energy.

Ronald Stein and Todd Royal are co-authors of the 2019 book, **Energy Made Easy** – *Helping Citizens become Energy Literate*. The book has received all 5-Star reviews on Amazon. com, just through the first six months of it release.

Ronald Stein, P.E. is an engineer and Founder of PTS Advance, drawing upon decades of project management and business development experiences. He is an internationally published columnist, and energy expert who writes frequently about all aspects of energy and economics; and is an energy policy advisor for The Heartland Institute.

Todd Royal is an independent public policy consultant in Los Angeles focusing on the geopolitical implications of energy. With hundreds of published Op Ed articles and his master's thesis titled, "Hydraulic Fracturing and the Revitalization of the American Economy," published in the U.S. Library of Congress in 2015, his scholarly works continue to focus on energy, geopolitics, national security, and foreign policy.

Ron and Todd believe it's dangerous and delusional to believe anything can be explained in sound bites, much less energy. They will discuss the pros and cons of renewables, coal, natural gas, oil, petroleum, and nuclear generated electricity, touted in the press and social media, to meet the needs of societies around the world.

Electricity and the thousands of products we use from petroleum derivatives are at the forefront of everything that touches our daily lives all over the world. Therefore, it is unwise to believe our global, interconnected world is going carbon-free, or wholeheartedly joining the get-off-the-fossil-fuel-crowd.

Any intellectual insight into energy facts without believing the green, or nothing-at-all movement – is anything other than political organizations attempting to elect officials who buy into taxpayer subsidies and write offs – when their shenanigans fail.

Just GREEN Electricity – *Helping Citizens Understand a World without Fossil Fuels* will make you look at energy and electricity in a fresh, new perspective. Ron and Todd believe this is desperately needed, and why they wrote this book. With the upcoming U.S. Presidential election, and global events taking place in China, Russia, Iran, Africa, India, and South America, explaining all parts of energy is needed more than ever.

Ron and Todd are energy agnostics who only want to share the facts for you to be further energy literate. Through their books, Op Ed articles, radio and TV interviews, and public speaking engagements, they provide in-depth discussions, and explanations in the book, **Just GREEN Electricity** – *Helping Citizens Understand a World without Fossil Fuels.* Chapters in the book will assist:

- *Attaining a better understanding about how the world has changed from societies that existed in primeval times, and the Dark Ages without electricity, transportation systems, or the thousands of products made from the petroleum derivatives that come from a barrel of crude oil.*
- *Be appreciative as to how the inventions of the automobile, airplane, and the use of petroleum in the early 1900's led us into the Industrial Revolution. Crude oil, natural gas, and coal changed – for the better – the lifestyles of every person living in developed countries such as, the U.S., Europe, Japan, South Korea, and Australia.*
- *Time to be cognizant about the dark side of electric vehicles, wind turbines, and solar panels that cannot function without exotic minerals being mined by children and slave labor in countries like the Congo in Africa, and China. The mere extraction of those exotic minerals used in the batteries of EVs and in the wind and solar systems presents social challenges, human rights abuses, and environmental degradations.*
- *Be mindful of how the world changed with the infrastructures of airlines, trains, vehicles, merchant ships, medications, fertilizers, cosmetics, and military equipment like aircraft carriers, battleships, planes, tanks and armor, trucks, troop carriers, and weaponry that did not exist before 1900. Everyone one of the pieces and parts of a thriving society and the world have exploded to meet the growing demands of society and the creative capabilities*

*of today's minds are demanding more and better prod-
ucts to enhance our lifestyles.*

- *Take notice as to why China and India, the two most
  populous countries in the world with billions of people,
  are rejecting renewables for coal fired power plants. How
  China and India use energy is obviously of great importance to world emissions levels, since coal is the dirtiest
  form of scalable, reliable, affordable, and abundant energy currently available to billions of people hungry for
  reliable electricity in the developing world.*

- *Awareness as to why Germany, Australia, and the United
  States are attempting to take on the leadership to convert
  the world from fossil fuels and all the derivatives that em-
  anate from petroleum to makes things and move products
  around the world. Closer to home, why does the state of
  California import electricity, and why is California the
  ONLY state in the continental USA that imports most of
  its crude oil needs from foreign countries?*

- *Learn about the climate alarmists who benefit from the
  all-electric narrative, since wind and solar electricity
  never accounts for the thousands of products that come
  from petroleum derivatives to "make things and move
  products".*

- *Understand how every part in wind turbines and solar
  panels comes from the derivatives from petroleum, and
  if the world's largest economy – the United States – shut
  down everything that caused emissions, including human
  life, emissions will still grow, because of China, India,
  and Africa. Literally, the U.S. could have 100 percent
  reductions in CO2 emissions, and global emissions will
  rise, because of China, India, and Africa, and their prolific usage of coal.*

By effectively communicating their in-depth research within their books, and Op Ed articles via in-person presentations and discussions, Ron and Todd desire to help students, non-governmental organizations, foundations, corporations, multi-national companies, government leaders, and all residents be better informed about all thing's energy related. Then energy literacy will grow about the many facets of renewable energies and fossil fuels that support lifestyles and economies around the world.

We look forward to hearing from you and speaking further about **Just GREEN Electricity** – *Helping Citizens Understand a World without Fossil Fuels.*

RONALD STEIN, P.E.
Founder and Ambassador for Energy & Infrastructure
Ronald.Stein@PTSadvance.com
Twitter: @PTSFounder

TODD ROYAL
Independent public policy consultant in Los Angeles focusing on the geopolitical implications of energy
www.ToddRoyalConsulting.WordPress.com
Twitter: @TCR_Consulting

INTRODUCTION

Wind and solar generated electricity can run motors, power on the lights, and charge all our electronic and communication gadgets; but electricity alone cannot make or produce any of the above-mentioned products. In fact, every part of wind turbines and solar panels comes from a barrel of crude. All the motors, lights, electronics, and communication equipment our modern world craves have their origins from crude oil. It is the most flexible product in the world, or possibly what has ever been invented or discovered. That dirty fossil fuel is societies' most in-demand product on planet earth.

If it weren't for the demands of society for airlines, cruise ships, transportation vehicles, and the thousands of products that are based on the chemicals derived from crude oil – there would be no need for the fossil fuels industry. The problem is not the supply, but the demands by society.

According to the British Petroleum (BP) Statistical Review of World Energy 2019, "global energy demand grew by 2.9 percent, faster than at any time since 2010-11," and projections into 2060 reveal oil, natural gas, coal, nuclear, renewables (solar and wind) will continue to grow.[1] Carbon emissions are also growing as demand for energy spikes upwards, with the biggest increase coming from Asia, or what has been called the "Asian Century." [2]

Keeping electricity flowing to homes, businesses, government, and militaries globally is a critical necessity for life, economic mobility, and vitality. We just left the best decade (2010-20) in human history for progress, innovation, and economic growth that led to hundreds of millions escaping poverty.[3] Electricity was at the forefront of these incredible, positive changes.

Literally, the entire developed world now runs off electricity. Electricity itself, however, is a grossly misunderstand commodity that has become the cure-all for society's demands, which have contributed to prosperity, and peace over war. What exactly is electricity? The U.S. Department of Energy defines electricity:

> "As the flow of electrical power – and it is a secondary energy source generated by the conversion of primary sources of energy like fossil fuels, nuclear, wind or solar."[4]

If you take away anything from this book, please understand that electricity is a stand-alone product, a secondary energy source. It is nothing without oil, petroleum; coal, natural gas, nuclear, biomass, ethanol, algae, solar panels, or wind turbines producing energy to electricity. Without energy from these sources you will never have electricity. Electricity is generated, and doesn't create any tangible products, or generate itself. Poverty and natural disasters overtake people, and nation's lives that do not have electricity, because it powers human flourishing and longevity.[5]

When there are over two billion people that do not have access to "reliable electricity" this should be the world's biggest issue, instead of focusing on earthly climate patterns and weather changes.[6] Why? Here are a few things Edison's invention has done and continues accomplishing for civilization – electricity can motor many forms of transportation, heat and air condition our homes, and businesses; and medical infrastructures that has people living longer than ever.[7]

Electricity alone will not solve the world's problems. War

will continue, famines will take place, and nations, and people groups will still perpetuate evil against one another. The most important fact about electricity under the best technology available is this – renewable electricity from solar panels and wind turbines is intermittent, mathematically unstable, and someone, somewhere needs to figure out how to build a better solar panel, and wind turbine for renewables to overtake oil, natural gas, coal, and nuclear for our global, electrical needs. [8]

The technologically advanced Germans have found that unstable electricity from solar panels and wind turbines have caused their electricity prices to be the highest in Europe, and made their electrical grid prone to blackouts.[9] Germany should solve the intermittent renewable problem, and stop worrying about climate change.[10]

Electricity from solar panels, and wind turbines cannot manufacture the products that make motors, lights, electrical and medical equipment as mentioned above, but it can power them. Most importantly for this book, and understanding electricity, it cannot make the over six thousand products that come from the derivatives of petroleum.[11]

What electricity brilliantly accomplishes is it keeps the lights on, the water flowing, and the machines in manufacturing facilities across our vast, interconnected world churning out products, goods and services to facilitate our luxurious, prosperous lifestyles.

Current climate alarmism believes electricity will solve global emission problems when the facts, and the problems with the climate movement's cult-like dogmatism, reveal a different outcome.[12] An outcome where demand for fossil fuels, energy, and electricity have never been higher in the United States.[13] Or China, where their coal consumption to meet electrical needs has quadrupled since 1975, and almost tripled from 1998-2013 according to the National Bureau of Statistics of China.[14]

Burning coal is known to release mercury, barium, selenium,

arsenic, and CO2 into the atmosphere if stack exhaust is not captured and mitigated; or attempting to use clean coal technology found in China, South Korea, Japan, Germany, Poland, and the U.S.[15] China's electrical hunger has caused them to increase its coal consumption from 1990 to 2015 from 1.05 billion tons to 3.97 billion tons. In 2016, coal made up sixty-two percent of China's energy use. Since 2011, China has consumed more coal than the rest of the world combined."[16]

India and China make up two out of seven people in the world, and India is also using more coal than ever before to deliver electricity to their growing population and economy.[17] A study published in Environmental Toxicology and Chemistry (ETC) highlights the "release of pollutants in one region (China, India, Asia as examples) can have implications beyond its border."[18]

Electricity does not have the ability to collaborate on environmental issues, stop, or even mitigate the coal-burning activities in India and China, or regions that have influenced environmental health in the United States. The ETC study suggested this is taking place.

Electricity connects us, and this electrical interconnectedness powered increasingly by coal, natural gas, and nuclear has a positive and negative human impact. Our hyper-civilized consumer patterns have driven global industrialization, and fostered some environmental controls, but at the center of all this its electricity.

The world's trying to change their current demands for transportation fuels, and products from fossil fuels to just electricity from wind and solar. The dilemma is those six thousand products that come from crude oil, manufactured in petrochemical facilities, which cannot run without electricity. Under current renewable technology, solar panels, and wind turbines, cannot manufacture these six thousand products, and their low energy density makes them a poor choice to meet global electricity needs for the various transportation infrastructures.[19]

The very products that make up the parts that capture the sun and wind from solar panels and wind turbines come from a barrel of crude oil.[20] The world needs to comprehend that less than half of a forty-two gallon barrel of crude oil is used to manufacture gasoline and diesel fuels used in all forms of transportation while the other half provides the derivatives that makes all the products we use in our daily lives.

Electricity is used at refineries to make the bulk of the "forty-five gallons of petroleum products," that come from forty-two gallons – with the entire process-taking place in technologically advanced refineries – based anywhere that has access to stable electricity; the refinery process gain is what allows forty-two-gallons to become forty-five-gallons.[21] Electricity then takes a major part of the barrel of oil to manufacture the chemicals in refineries, and petrochemical plants for everyday products such as makeup, plastics, and medications.

One of the biggest areas electricity makes our lives better is in the transportation sector by delivering fuels that fly approximately 39,000 planes that move over four billion passengers a year.[22] The sixty thousand merchant ships in commercial maritime transport delivers the six thousand products from a barrel of crude oil into ports and shipping lanes from South Africa to Iceland, and we see that ninety percent of the world's goods cannot be delivered without reliable, and stable electricity.[23]

It's important to note that crude oil is not used for electrical generation, only coal, natural gas, and nuclear fulfill that role under the auspices of energy being scalable, affordable, abundant, reliable, and flexible. Renewables (sun and wind) while sustainable, are intermittent, and are not affordable, reliable, flexible or scalable.[24] Nuclear is the only wide-scale source of carbon-free electrical generation currently.

Emerging countries that have been the epicenter of human wretchedness, such as India, are demanding fossil fuels like never before. While the west, led by the United States, the European Union (EU), and United Nations are attempting vast societal changes by ridding the world of fossil fuels. Then how do we replace the thousands of products, millions of flights, and the need to move people and products to all corners and continents on the globe that are accomplished by fossil fuels?

We've had almost 200 years to develop clones or generics to replace the products we get from crude oil. Today, no one has that answer at this time, and current technological constraints have not come up with a way to make solar panels and wind turbines have the ability to withdraw from fossil fuels, or eliminate the products humanity can't live without like the iPhone or clean water. Most of all electricity cannot accomplish eradicating fossil fuels.

Relying on just electricity is unlikely. All the mineral products and metals needed to make wind turbines and solar panels rely on worldwide mining and transportation equipment that are made with products from fossil fuels and powered by fuels manufactured from crude oil.

Only reducing the social demands, the industrial revolution brought and returning to the early 1800's can accomplish life without fossil fuels. Reducing demand for modern life, and the supply dwindles back to the time when wind, sun, and oxen were the prime movers of society.

Simply changing vehicles from fossil fuel energy to battery powered vehicles (EVs), or somehow bringing nuclear and military vessel technology into the consumer, business and government transportation economy would bring low-carbon, or possibly zero-carbon economies that have become all the rage.

The facts state a low-carbon, or zero-carbon society is impossible for electricity, or global economies for decades ahead under current technological constraints.[25] Electricity is unable to mitigate any of these transportation, economic, or military preparedness issues at this time, or in the near future.

There is a deep quandary taking place in the world of electricity, because the U.S. Department of Energy's, Energy

Information Administration (EIA), published in late September 2019, its International Energy Outlook 2019 (IEO2019) that said, "the world's energy consumption is set to increase by almost fifty percent until 2050."[26]

This study echoed the same thing the BP *Statistical Review of World Energy* said on the first page. The world needs energy, and it will come from fossil fuels, nuclear, and renewables. Nowhere does either study say that electricity can power itself.

The U.S., EU, and other western-leaning nations may want a "Green New Deal," where electricity runs the world, but what this book will attempt to show is factually there is more to energy than electricity. The largest misconception about electricity in vogue is renewables can electrify the world, and that simply isn't true.[27]

There still isn't a cost-effective way, nor the energy density, for solar panels, and wind turbines used for electricity to power the U.S., EU, or any modern government, military, country, nation, continent, or society.[28] Electricity needs fossil fuels and nuclear energy to provide continuous and uninterruptible electricity more than ever.[29]

What we'll term environmental electricity still doesn't mean that electricity can fulfill the goals of the climate change/global warming movement (CCGW). The technology still isn't there for electricity to only come from solar panels, wind turbines, biomass, hydropower, or algae without continuous back up from coal, natural gas, or nuclear.[30]

Multinational firms may commit to emission targets, and say climate change is the issue, but none of them can definitively say when pressed, how they would only use electricity to run their international companies to alleviate climate change?[31] More importantly, these firms don't understand that climate apocalyptic predictions have been wrong for over fifty years.[32]

The question(s) for environmental-electricity advocates is whether mankind is causing it based on historical eras when it was hotter and cooler than our current climate? No one seems to have the answer for that question.[33]

Making dire climate change predictions is morally wrong, and does nothing to address how two billion people will get reliable electricity? Groups like Extinction Rebellion do vast harm to billions of people by turning the environment into "voodoo science," with zero understanding of environmental stewardship, electrical generation, and human longevity.[34]

Thank goodness for an environmentalist such as Michael Shellenberger, *Time Magazine's*, "Hero of the Environment" who wrote in 2019, "Why Apocalyptic Claims About Climate Change Are Wrong;" Mr. Shellenberger boldly wrote:

> "I also care about getting the facts and science right and have in recent months (Fall 2019) corrected inaccurate and apocalyptic news media coverage of fires in the Amazon[35] and fires in California,[36] both of which have been improperly presented as resulting primarily from climate change."[37]

For all of electricity's glory to mankind it cannot mitigate the environmental-electricity movement's use of a sixteen year old, diagnosed with Asperger's, high-functioning autism, and obsessive-compulsive disorder, crisis-mad, high school dropout, used for the non-profit industrial complex's glorification, little girl – Greta Thunberg – as the new spokesperson for CCGW.[38]

The CCGW movement has officially lost credibility, and no longer able to discern what electricity can, and cannot accomplish when a late 2019 study was published in Obesity, the online magazine of The Obesity Society stating, "obesity associated with greater greenhouse gas emissions."[39] Add a manipulated child into the mix, and the very solutions for electricity that Greta is being used for to promote would devastate the poor all over the world.[40]

As fathers, nothing troubles us more than watching this manipulated little girl, who should be back in school, be used to promote an agenda.[41] We are living in "unserious times," when the geopolitics of electricity is cast aside to seriously ponder the rage-filled antics of a child, or believe that overweight people are causing the earth to warm, cool, or somewhere in-between.[42]

The most troubling part about a child being taken seriously on global environmental issues, are the reasons why two billion people are denied the ability for stable electricity. Poor people, nations, and continents – like Africa that have over 600 million people without electricity – should be given abundant, inexpensive, sources of energy provided by coal, petroleum, oil, natural gas, and nuclear.[43] MIT climate scientist Kerry Emanuel says:

> "If you want to minimize carbon dioxide in the atmosphere in 2070 you might want to accelerate the burning of coal in India today (as an example). It doesn't sound like it makes sense. Coal is terrible for carbon. But it's by burning a lot of coal that they (India) make themselves wealthier, (similar to America in the early 1900's), and by making themselves wealthier (any poor region of the world) they have fewer children, and you don't have as many people burning carbon, you might be better off in 2070 We shouldn't be forced to choose between lifting people out of poverty and doing something for the climate."

In September 2019 the Iranian-Saudi Arabian rivalry reached a new, dangerous peak when the Iranians using the Houthi's, an Islamic Shia militant group affiliated with Iran, attacked the Abqaig oil facility in Saudi Arabia using sophisticated drone strikes.[44]

This could have been the greatest disruption of oil and natural gas to electricity supplies since the 1991 Gulf War. Middle Eastern volatility has been taking place for thousands of years, and will likely continue, but if electrical blackouts happen over this volatility then chaos ensues.[45]

Electricity has zero answers for Greta Thunberg, politicized climate change, lifting people out of wretched poverty, or the most significant geopolitical event concerning electricity of the last forty years that happened in the Middle East in 2019.

That wasn't the case a hundred years ago. Electricity could have been shut down across continents or seen price spikes. Then add the collapse of Venezuelan oil supplies, over embracing socialism, and the world should have seen electricity blackouts on all seven continents.[46]

Instead, the abundance of oil in the markets from U.S. production "prevented panic and allowed the markets to move on with a minimal disruption."[47] Global energy supplies – particularly oil, natural, and coal – remain solid with minimal price disruption. The biggest reasons this is taking place is U.S. shale oil, fracking, and natural gas production.

For the first time since 1949 when energy records starting being kept, the U.S. is now a net exporter.[48] The EIA "predicts the net crude oil exports to continue exceeding net imports by an average of at least eighty thousand barrel per day (bpd) in 2020."[49] This is great for electricity deliverability and reliability, because electricity doesn't create itself, it needs oil, natural gas, coal, nuclear, biomass, sunshine, or wind. With bountiful supplies from the U.S. inserted into global energy, and electrical markets, therefore the world diverted an electrical and geopolitical crisis.[50]

However, the potential for a considerable threat to electrical deliverability on the horizon this decade 2020-2030 is the Russian Nord Stream 2 pipeline, which brings the largest delivery of natural gas to electricity to the European Union.[51]

These activist, and geopolitical phenomenal situations are outside the realm, and scope of electricity's expertise. Now add the corruption of global scientific societies more interested in chasing taxpayer monies to promote climate change, and you can understand why countries never leave poverty, and billions do not have access to better lives for their countries, families, and children.[52] Whether man is, or isn't causing climate change, electricity is a stand-alone entity that cannot function without abundant, reliable, stable, affordable, and flexible oil, natural gas, petroleum, coal, and nuclear generated electricity.

What this book will attempt to show is that electricity isn't the balm for our exploding electrical needs and demands. Forecasts, green new deals, and informed geopolitical discussions are important factors that could derail electricity, but only fossil fuels and nuclear can withstand these disruptions.

Electricity needs both fossil fuels and nuclear to function, and in the context of Brexit, the 2020 U.S. Presidential election, instability in the Middle East, and the U.S.-China trade war/tensions are circumstances electricity faces. This book will try to explain them all in a clear, concise, well-researched and documented way, while showing the deep ramifications of a world without fossil fuels, nuclear power, and the products from petroleum derivatives that support lifestyles and global security.

As an added benefit to you, the reader, each chapter is a stand-alone read on energy subjects. You may not be interested in the entire spectrum of electricity, but can selectively pick, and choose hot energy topics trending on current news coverage, or social media sites.

We hope you enjoy the book, learn a little more about electricity than you knew previously, and ponder the unanticipated consequences of a world without fossil fuel, and nuclear generated electricity.

Thank you for reading our book.

RON & TODD

CHAPTER ONE

THE GREEN NEW DEAL FUTURE
The World without Fossil Fuels

By Ronald Stein

INTRODUCTION

Before the Great Horse Manure Crisis of 1894, the world was without fossil fuels.[53]

Figure 1-1 By the late nineteenth century, the dawn of the industrial age had presented an interesting problem for large cities all around the world—they were "drowning in horse manure." The industrial age meant that people were flocking to cities in large numbers, and the nature of industrialization created a higher demand for the transport of both people and goods—all meaning more horses.

In 1900, there were over eleven thousand hansom cabs on the streets of London alone. There were also several thousand horse-drawn buses, each needing twelve horses per day, making a total of over fifty thousand horses transporting people around the city each day. To add to this, there were yet more horse-drawn carts and drays delivering goods around what was then the largest city in the world.

"In 50 years, every street in London will be buried under nine feet of manure."
Times of London, 1894

A lesson from "The Great Horse Manure Crisis of 1894"

Figure 1-2: A lesson from the Great Horse Manure Crisis of 1894, London ca. 1894[54]

On average, a horse will produce between fifteen and thirty-five pounds of manure and two pints of urine per day. The average life expectancy of the hardworking horses was around three years, meaning that they were being laid to rest on the street in growing numbers. The bodies were often left to putrefy so the corpses could be more easily cut into multiple pieces for removal. The health risks these decaying horses created were compounded by flies, insects, and other pests, which spread typhoid fever and other diseases.

The streets of London were literally a cesspit of disease that was contributing to the decay of the city and those who lived there.

Climate change advocates, and the signatories to the Green New Deal (GND) and Paris Agreement agree the world needs to abandon fossil fuels and focus on renewable, intermittent electricity to save the world from itself! Their plans would return us to London in 1894, saturated in horse dung and urine.

WHAT IS THE GREEN NEW DEAL RESOLUTION?[55]

Senator Edward Markey and Representative Alexandria Ocasio-Cortez released a fourteen-page resolution for their Green New Deal on February 7, 2019. The approach pushes for transitioning the United States to 100 percent renewable, zero-emission electricity sources such as wind and solar power. This also includes investments into electric cars and high-speed rail systems. This factors in the "social cost of carbon" that had been part of the Obama administration's plans for addressing climate change within ten years. The resolution calls for a "10-year national mobilization" with twelve primary goals:

- Eliminating all electricity generated by coal, natural gas, and nuclear power.
- Meeting 100 percent of the electricity demands in the United States through clean, renewable, and zero-emission electricity sources.
- Eliminating internal combustion engine vehicles.
- Replacing internal combustion vehicles with those that run on renewable generated electricity.
- Upgrading all existing buildings in the United States and building new buildings to achieve maximal energy efficiency, water efficiency, safety, affordability, comfort, and durability, including through electrification.
- Replacing air travel with electrified high-speed rail systems.
- Overhauling transportation systems in the United States to eliminate pollution and greenhouse gas emissions from the transportation sector as much as is technologically feasible, including through investment in (i) zero-emission vehicle infrastructure and manufacturing; (ii) clean, affordable, and accessible public transportation; and (iii) high-speed rail.

- Guaranteeing a job with a family-sustaining wage, adequate family and medical leave, paid vacations, and retirement security to all people of the United States.
- Providing all people of the United States with (i) high-quality health care; (ii) affordable, safe, and adequate housing; (iii) economic security; and (iv) access to clean water, clean air, healthy and affordable food, and nature.
- Building or upgrading to energy-efficient, distributed, and "smart" power grids, and working to ensure affordable access to electricity.
- Spurring massive growth in clean manufacturing in the United States and removing pollution and greenhouse gas emissions from manufacturing and industry as much as is technologically feasible.
- Working collaboratively with farmers and ranchers in the United States to eliminate pollution and greenhouse gas emissions from the agricultural sector as much as is technologically feasible.

SPECIAL NOTE

Senator Bernie Sanders's version of the GND goes well beyond the GND presented by Senator Edward Markey and Representative Alexandria Ocasio-Cortez.[56] In addition to sunsetting the entire fossil fuel industry in America, Sanders's plan would:

- Ban imports and exports of fossil fuels. Congress' decision in 2015 to lift the ban on exporting fossil fuels was a mistake. We must no longer export any fossil fuels. Our coal and natural gas are contributing to increased emissions abroad. We will also end the importation of fossil fuels to end incentives for extraction around the world.

We can meet our energy needs and ensure energy security and independence without these imports.

That one action item of Bernie's GND plan to ban imports of fossil fuels from other countries will ground Air Force One, close all American airports, and eliminate America's military to name a few consequences of his plan.

THE PROS OF THE GND[57]

- *On the surface, the Green New Deal (GND) sounds enlightening. Use a nonexistent super grid of renewable, intermittent electricity to replace fossil fuels so we can all breath air with no emissions from energy production.*
- *Politicians both here and abroad are supportive of the GND to sunset the oil industry.*
- *The GND benefits from short, emotional social media tweets from Alexandria Ocasio-Cortez (AOC) and her 5.9 million followers, Greta Thunberg and her 3 million followers, Al Gore's 3.1 million followers, Tom Steyer's 250 thousand followers, and Jane Fonda's 500 thousand followers, all bumbling about the doomsday that's coming.*
- *Many states and localities have taken significant actions on climate change, renewable energy, energy conservation and efficiency, and sustainable economic development. Some advocates favor advancing zero-emission nuclear power in this mix and others do not.*

THE CONS OF THE GND[58]

- *By sunsetting the fossil fuels industry, the Green New Deal (GND) and Paris Agreement would also sunset the renewable industry that's supposed to be the salvation for the world, as there would be no components to build the turbines and solar panels with.*

- *Ocasio-Cortez's Green New Plan (GNP) is most likely unaffordable even for "rich" nations, making it an irresponsible plan to run on every economic platform across the globe. Many of the world's inhabitants survive in deplorable economic and living conditions where the GNP cannot work.*

- *The GND and Paris Agreement do not discuss the questionable and nontransparent labor conditions and loose environmental regulations at the sites around the world where the products and metals required for renewables are produced.*

- *Mining for the sixteen components needed to build wind turbines: aggregates and crushed stone (for concrete), bauxite (aluminum), clay and shale (cement), coal, cobalt (magnets), copper (wiring), gypsum (cement), iron ore (steel), limestone, molybdenum (alloy in steel), rare earths (magnets; batteries), sand and gravel (cement and concrete), and zinc (galvanizing).*

- *Mining for the seventeen components needed to build solar panels: arsenic (gallium-arsenide semiconductor chips), bauxite (aluminum), boron minerals, cadmium (thin film solar cells), coal (by-product coke is used to make steel), copper (wiring; thin film solar cells), gallium (solar cells), indium (solar cells), iron ore (steel), molybdenum (photovoltaic cells), lead (batteries), phosphate rock (phosphorous), selenium (solar cells), silica (solar cells), silver (solar cells), tellurium (solar cells), and titanium dioxide (solar panels).*

- *The products for wind and solar are mined in more than sixty countries throughout the world, including Algeria, Arabia, Argentina, Armenia, Australia, Belgium, Bolivia, Brazil, Canada, Chile, China, Congo*

(Kinshasa), Cuba, Egypt, Finland, France, Germany, Greece, Guinea, Guyana, India, Indonesia, Iran, Ireland, Italy, Jamaica, Japan, Kazakhstan, Madagascar, Malaysia, Mexico, Mongolia, Morocco, Mozambique, New Caledonia, Oman, Pakistan, Papua New Guinea, Peru, the Philippines, Poland, Russia, Saudi Arabia, Sierra Leone, Slovakia, South Africa, South Korea, Spain, Suriname, Sweden, Thailand, Turkey, Ukraine, the United Kingdom, the United States, Uzbekistan, Venezuela, Vietnam, Western Sahara, and Zambia.

- *The Green New Deal is oblivious to the unintended consequences of a world without fossil fuels. The signatories have failed to imagine how life was without that industry just a few hundred years ago before 1900, when we had no military equipment like we have today; no cell phones, computers, and iPads; no vehicles; no airlines, which now move four billion people around the world annually; no cruise ships, which now move twenty-five million passengers around the world annually; no merchant ships, which move billions of dollars of products monthly throughout the world; no tires for vehicles or no asphalt for roads; no water filtration systems; no sanitation systems; no space program; no medications or medical equipment; no vaccines; no fertilizers to help feed billions; and no pesticides to control locusts and other pests.*

- *Under the Green New Deal, we would have electricity available, but nothing to power, since virtually everything we have today is made with the chemicals and by-products manufactured from crude oil.*

- *The oil industry currently runs this world's economy. If we remove this industry before we have an alternative industry to replace it, we will be embarking on an unknown road and possibly suffer severe economic consequences.*

- *Modern industries are increasing, not decreasing, their need for deep earth minerals/fuels to "make products and move things" each year.*
- *The GND is so expensive; the cost would exceed fifty trillion dollars for the first decade and cost a total of ninety-three trillion dollars to achieve all its goals.[59]*
- *Unanswered questions of how to finance the GND's guaranteed jobs for everyone with no infrastructures to work at, and high-quality healthcare for all with no medications or medical equipment.*
- *According to a report from the International Energy Agency[60] (IEA), solar power could very well be responsible for over a third of our energy supply by the year 2060.*
- *With their prolific usage of coal, Chinese and Indian emissions are now higher than the U.S. and Europe combined and increasing rapidly as hundreds of millions of people start to climb out of energy poverty.*
- *Chinese and Indian emissions are now higher than the U.S. and Europe combined and increasing rapidly as hundreds of millions of people start to climb out of energy poverty.*

WHAT IS THE PARIS AGREEMENT?[61]

At the COP 21 in Paris, on 12 December 2015, parties to the United Nations Framework Convention on Climate Change (UNFCCC) reached a landmark agreement to combat climate change, and intensify actions and investments needed for a sustainable low carbon future.

The Paris climate agreement is an initiative run by the UNFCCC to help reduce global warming. The goal is to keep the Earth's temperature from rising more than 2 degrees Celsius by curbing the amount of greenhouse gas emissions allowed by each participating nation.

THE PRO'S OF THE PARIS ACCORD[62]

- *The "leaders" of the 184 ratifying countries of the Paris Agreement wish to delay and eventually stop any further exploration of deep earth minerals by diverting those countries' efforts toward intermittent renewables of Industrial wind and industrial solar for electricity.*
- *The Paris Agreement has global support from 184 ratifying countries.*
- *The Paris agreement is proof that climate change has become a priority for much of the world.*
- *The nations that have signed the agreement are signaling a world-wide commitment and investment into more intermittent electricity from wind and solar.*

THE CON'S OF THE PARIS ACCORD[63]

- *The Paris Agreement excludes Russia, Turkey and Iran as those countries believe the countries that control natural gas and oil will control the world. Putin of Russia understands WW I and II were both won with fossil fuels to move planes, ships, tanks, troops and supplies.*
- *The "leaders" of the get-off-fossil fuels are indirectly supporting Russia's efforts to have all the ratifying countries of the Paris Agreement delay or stop any further exploration of fossil fuels by diverting those countries efforts toward intermittent renewables of Industrial wind and solar for electricity. Russia, knows whoever controls oil, controls the world.*
- *Those "leaders" of the get-off-oil campaign seem to be oblivious to the fact that 100 percent of the industries*

*that use fossil fuels to "make products and move things"
to support the economies around the world are increasing
their usage each year, not decreasing it. The world has become accustomed to the lifestyles provided by elaborate
infrastructures and military that support the prosperity
of growing populations that are all based on fossil fuels.*

- *With the Paris Agreement in place, nations around the
world are defying the Energy Industry Administration
(EIA) projections that world energy production and consumption continues to grow for all fossil fuels of coal,
oil and gas, and that most electricity in the world will be
generated by sources other than renewables. i.e., Nuclear,
Natural Gas, and Coal.*
- *The Paris Agreement Has Different Rules for Different
Countries*

Here are all the things that electricity can do for civilization:

- Provide electricity to run the motors of vehicles, heating,
air conditioners.
- Provide electricity for lighting
- Provide electricity for electronics
- Provide electricity for the medical infrastructure

Basically, electricity can power the motors, lights and electronics, but it can't make the motors, lights and electronics!

Lets' be clear about what that means. First, it's not renewable
energy, it's only renewable electricity, and more accurately its
only intermittent electricity. Renewables have been the primary
driver for residents of Germany, Australia, and California behind
the high costs of electricity, as renewables have proven to be an
inefficient redundant source of electricity to the continuous uninterruptible electricity from coal, natural gas, and nuclear. Second
and most important is, electricity alone is unable to support

militaries, aviation, and merchant ships, and all the transportation infrastructure that support commerce.

Russia uses its vast energy resources[64] exactly the way the
Soviet Union used its nuclear weapons in the Cold War, to control nations and scare global citizens.[65] Russia now has the ability
"to increasingly wield oil as a geopolitical tool"[66], spreading its
influence around the world, and challenging the interests of the
United States.[67] Geopolitics has been defined as "How a country's geography and location affect its foreign, economic (think
energy) and military policy."[68] Vladimir Putin's Russia is challenging Western Democratic organizations like NATO and the
EU by attacking them through their dependence on Russian oil
and natural gas.[69]

The "Praise the Lord" (PTL) empire that preacher Jim Bakker
built with wife Tammy crumbled from scandals thirty years
ago[70]. Today, it seems like we're being mesmerized again in the
press and social media with the rhetoric about the need for industrial wind and solar generated electricity to save the world
from human destruction.

Everyone knows that electricity is used extensively in the
residential, commercial, transportation, and the military sectors,
to power motors and lights; but it's the six thousand products
that get manufactured from petroleum derivatives that are used
to make those motors, lights, and electronics. Noticeable by
their absence, from turbines and solar panels, are those crude oil
chemicals that renewables are currently incapable of providing[71].

We've had almost 200 years to develop clones or generics
to replace the products we get from crude oil such as: medications, electronics, communications, tires, asphalt, fertilizers,
military and transportation equipment. The social needs of our
materialistic societies are most likely going to remain for continuous, uninterruptable, and reliable electricity from coal or
natural gas generation backup, and for all those chemicals that

get manufactured out of crude oil, that makes everything else that's part of our daily lifestyles.

Energy storage could revolutionize industries in the next ten years, but despite the preaching's about these renewable saviors, it's becoming obvious that due to their intermittency and unreliability, and their inability to replace any of the chemicals from crude oil that account for the all the products in our daily lives, societies around the world may not be too thrilled about the needed social changes to live on just electricity[72].

All the mineral products and metals[73] needed to make wind turbines and solar panels rely on worldwide mining and transportation equipment that are made with the products from fossil fuels and powered by the fuels manufactured from crude oil.

The same critical minerals are also used for other technologies for the Clean-Energy Transition

Clean-Energy Technologies	Material of Concern	Uses in the Technology
Electric Vehicles	Co	Li-ion batteries (electrodes)
	Li	Li-ion batteries (electrolytes)
	Dy and Nd	Magnets in electric motors
	Ag	Electrical interconnects
Solar PV	Ga, Ge, In, Te, and Se	Various thin0film PV layers
	La	SOFC electrodes and electrolytes
Fuel cells and FCV's	Pt	PEMFC electrodes
Wind turbines	Dy and Nd	Permanent magnets in generators

Abbreviations are as follows: FCV, fuel cell vehicles; PEMFC, polymer electrolyte membrane fuel cell; PV, photovoltaics; SOFC, solid-oxide fuel cell.

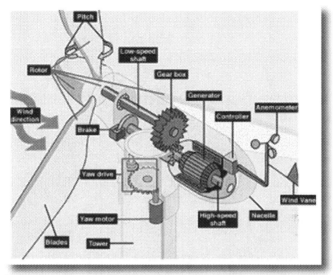

Figure 1-3: Parts of a wind turbine[74] are the tower, blades, shaft, gear box, controller, generator, brake. anemometer, yaw drive and motor, and the nacelle which is a cover housing that houses all the generating components in a wind turbine, including the generator, gearbox, drive train, and brake assembly.

Parts of solar systems[75] are the solar panels, solar array mounting racks, array DC disconnect, inverter, battery pack, power meter, and circuit breaker panel.

Yes, rare earth minerals are required for batteries. The Chinese have bought up most of the known rare earth deposits in the world[76]. Extraction of these minerals will do more harm to the environment than the burning of fossil fuels.

What folks seem to forget is that it takes fossil fuels to make and operate the equipment to extract and refine the rare earths to make the batteries. And, unfortunately, batteries do not last forever as most of us who drive cars are aware. So, converting to battery power is not a one-shot deal but on ongoing proposition. I would not want to have to rely on battery power as I know their life spans are limited. Changing those "AA" batteries is easy, but constantly changing those "big" EV batteries in incomprehensible.

Environmental degradation is an integral part of the renewables process. Bear in mind, the underlying motive of this survey is a profound wish for renewables to work. But without examining the negatives associated with renewables, and the need to rid the sector of predatory faux-environmentalists, and opportunistic investors, this form of clean electricity will most likely never become a dominant force globally. The rare earth materials needed for renewables are not unlike those needed for electric vehicles (EVs), namely cobalt and lithium, which come from the Democratic Republic of Congo where corruption, environmental pollution, extreme poverty, and child labor conditions are the norm[77].

Never discussed by the GND, or Paris Accord sponsors are the questionable and non-transparent labor conditions and loose environmental regulations at the sites around the world for the products and metals required for renewables[78].

- A list of the sixteen components needed to build wind turbines are: Aggregates and Crushed Stone (for concrete), Bauxite (aluminum, Clay and Shale (for cement), Coal, Cobalt (magnets), Copper (wiring), Gypsum (for cement), Iron ore (steel), Limestone, Molybdenum (alloy in steel), Rare Earths (magnets; batteries), Sand and Gravel (for cement and concrete), and Zinc (galvanizing).
- A list of the seventeen components needed to build solar panels are: Arsenic (gallium-arsenide semiconductor chips), Bauxite (aluminum), Boron Minerals, Cadmium (thin film solar cells), Coal (by-product coke is used to make steel), Copper (wiring; thin film solar cells), Gallium (solar cells), Indium (solar cells), Iron ore (steel), Molybdenum (photovoltaic cells), Lead (batteries), Phosphate rock (phosphorous), Selenium (solar cells), Silica (solar cells), Silver (solar cells), Tellurium (solar cells), and Titanium dioxide (solar panels).

- The origins of the products for wind and solar are mined throughout the world, inclusive of more than 60 countries of Algeria, Arabia, Argentina, Armenia, Australia, Belgium, Bolivia, Brazil, Canada, Chile, China, Congo (Kinshasa), Cuba, Egypt, Finland, France, Germany, Greece, Guinea, Guyana, India, Indonesia, Iran, Ireland, Italy, Jamaica, Japan, Kazakhstan, Madagascar, Malaysia, Mexico, Mongolia, Morocco, Mozambique, New Caledonia, Oman, Pakistan, Papua New Guinea, Peru, Philippines, Poland, Republic of Korea, Russia, Saudi Arabia, Sierra Leone, Slovakia, South Africa, Spain, Suriname, Sweden, Thailand, Turkey, Ukraine, United Kingdom, United States, Uzbekistan, Venezuela, Vietnam, Western Sahara, and Zambia.

The hype these days is to stop using those dirty fossil fuel driven cars and trucks and convert everyone to those clean electric vehicles. But wait!

Before you jump onto the EV train, those EV's have a very dark side of environmental atrocities and a non-existing transparency of human rights abuses associated with mining for the exotic minerals that power the EV's and iPhones.

The key minerals used in today's batteries are cobalt, of which sixty percent is sourced from one country, the Democratic Republic of the Congo[79] (DRC), and lithium, of which more than fifty percent is sourced from the Lithium Triangle in South America[80], which covers parts of Argentina, Bolivia and Chile. Today twenty percent of cobalt is mined by hand. Amnesty International has documented children and adults mining cobalt[81] in narrow man-made tunnels, at risk of fatal accidents and serious lung disease.

The exotic minerals of lithium and cobalt are both extremely limited in their supply and available locations, compared to crude oil that can be found in almost every country and ocean and at various depths. The limitations of supply and minable locations for these in-demand commodities present a very serious challenge as to how to continue the EV revolution when those supplies begin to diminish.

Rising tensions between the United States and the countries suppling those rare earths for leverage may lead to trade wars[82] between the global economic powers and result in a national security risk for the American economy.

The mere extraction of the exotic minerals cobalt and lithium[83] used in the batteries of EVs energy battery storage systems present social challenges, human rights abuses, and environmental degradation.[84] Not only are working conditions hazardous, but living conditions are abysmal, with workers making such meager wages that they are forced to live in abject poverty; and, whether on-duty or off, regularly exposed to out-of-control pollution and many other environmental issues that cannot be ignored.

The cobalt mined by children and adults in these horrendous conditions in the DRC in Africa then enters the supply chains of some of the world's biggest brands. There are no known "clean" supply chains for lithium and cobalt, yet the richest and most powerful companies in the world continue to offer up the most complex and implausible excuses for not investigating their own supply chains.

Figure 1-4 Children and adults mining for exotic minerals for "clean" electricity[85]

The mining of exotic minerals around the world may be like the conflict of "blood" diamonds[86] that originate from areas controlled by forces or factions opposed to legitimate and internationally recognized governments.

Tesla Motors' "dirty little secret"[87] is turning into a major problem for the EV industry—and perhaps mankind. If you think Tesla's Model S is the green car of the future, think again. The promises of energy independence, a reduction in greenhouse gas emissions, and lower fuel costs, are all factors behind the rise in the popularity of electric vehicles. Unfortunately, under scrutiny, all these promises prove to be more fiction than fact.

Recently, the Environmental Protection Agency and the U.S. Department of Energy undertook a study to look at the environmental impact[88] of lithium-ion batteries for EVs. The study showed that batteries that use cathodes with nickel and cobalt, as well as solvent-based electrode processing, have the highest potential for environmental impacts, including resource depletion, global warming, ecological toxicity, and adverse effects on human health.

The largest contributing processes include those associated with the production, processing, and use of cobalt and nickel metal compounds, which may cause adverse respiratory, pulmonary, and neurological effects to those exposed.

As demand for rechargeable batteries grows, companies have a responsibility to prove that they have ethical supply chains, a priority when implementing green policies, and are not profiting from the misery of miners working in terrible conditions like those in the DRC. The energy solutions of the future of EV's, solar panels, and wind turbines must not be built on human rights abuses[89].

When a company has contributed to, or benefited from, child labor or adults working in hazardous conditions, it has a responsibility to remediate the harm suffered. This means working with other companies and governments to remove children from the worst forms of child labor and support their reintegration into schools, as well as addressing health and psychological needs.

Non-existent proactive environmental regulations and human rights abuses[90] are both on the dark side of green technology.

All mining and processing activities for renewables[91] to get the iron ore and other metals that go into turbine manufacturing, transporting the huge blade beasts to the sites, and decommissioning them, are all energy intensive activities that generally rely on fossil fuels and the products from crude oil and leave difficult wastes behind to dispose of during decommissioning.

Unbeknownst to the signatories of the Green New Deal (GND) and Paris Accord desirous of sunsetting the fossil fuels industry for a world surviving on renewable electricity would also sunset its own renewable industry, as there would be no components to build the turbines and panels!

The useful life of wind turbines is limited, generally from fifteen to twenty years, but none of the decommissioning plans are public. Mining projects, oil production sites, and nuclear generation sites are required to provide for decommissioning and restoration details down to the last dandelion. Would governments and the Green movement allow a decommissioned mine, oil or nuclear site of similar latitude and size?

Wind power kills birds and bats. Environmental groups have long been concerned with wind turbines killing bats and birds, including many protected endangered species. Wind turbines alone are estimated to kill 600 thousand birds and a million bats annually.

In 2017, the former President Obama's administration finalized a rule that lets wind-energy companies operate high-speed turbines for up to thirty years — even if it means killing or injuring thousands of species protected under the Bald and Golden Eagle Protection Act and the Migratory Bird Treaty Act.

Under the new rule, industrial wind may acquire an eagle "take" permit from the U.S. Fish and Wildlife Service (USFWS) that allows the site to participate in the nationwide killing of up to 4,200 bald eagles annually[92], under incidental "take" permits without compensatory mitigation. It's shocking that industrial wind can legally obtain permits from the USFWS to kill those majestic bald eagles. We cry foul!

We wonder if the renewable industry is proud of those new jobs being created also include those that need to clean up the mess from those creatures chopped up by the industrial wind generator blades and from those fried from the heat from the industrial solar panels?

GND mandated solar and wind facilities would need to be located further from populated urban areas than natural gas, coal, and nuclear facilities, requiring a major expansion of high voltage transmission lines. But as recent wildfires in California show, such lines can cause major environmental damage. Environmental groups have opposed such lines and, when they exist, have opposed the clearing of underbrush, fallen trees, and the like as "unnatural" thus making devastating fires more probable as the fire "fuel" continues to grow.

Extensive pollution occurs from the material and processes to support wind and solar. The steel, concrete, rare earth minerals, and other materials necessary to build enough wind turbines and solar panels to power the nation would require mining and production on a scale most environmentalists strongly oppose.

Numerous American states and several countries, notably Germany, are experiencing major environmental problems due to the disposal of solar panels, wind turbines, and batteries after they lose their usefulness Again, this environmental harm is ignored by GND advocates and would multiply substantially under the GND plan.

We can be preached to forever about "clean electricity" messages, and bedazzle farmers with the prospects of on-going revenue from renewables, but the extensive mining worldwide for turbine and solar materials, and the decommissioning details, and the social changes that would be necessitated without the thousands of products from those deep earth minerals and fuels, remain the dark side of the unspoken realities of renewables.

GERMANY

- *Germany tried to step up as a leader on climate change, by phasing out nuclear, and pioneered a system of subsidies for wind and solar that sparked a global boom in manufacturing those technologies.*

- *Germany was the first major economy to make a big shift in its energy mix toward low carbon sources, but Germany is failing to meet its climate goals of reducing carbon-dioxide emissions[93] even after spending over $580 billion by 2025 to overhaul its energy systems. Germany's emissions miss should be a "wake-up call" for governments everywhere.*

- *Like Germany, America's renewables are becoming an increasing share in electricity generation, but at a HIGH COST. The emission reduction goals have increased the costs of electricity and transportation fuels and may be very contributory to America's growing homelessness and poverty populations.*

- *Power prices in Germany are among the highest in Europe[94] – but many customers continue to support the switch to renewable energy sources regardless. Today, German households pay almost 50 percent more for electricity than they did in 2006. Much of that increase in electricity cost is the Renewable Surcharge that has increased over the same period by 770 percent.*

- *Even with the financial and environmental disastrous results in Germany over the last few decades, German Chancellor Angela Merkel vowed in 2019[95] that her nation would do "everything humanly possible" to curb the impacts of climate change. She isn't giving up and is willing to sacrifice the Germany economy to pursue her personal believes.*

America is about to take one giant step toward following Germany's failed climate goals which should be a wake-up all for governments everywhere[96], but it appears that America, from California to New York, wants to follow the German failure.

U.S. Senator Chris Coons and Senator Dianne Feinstein recently announced that they will introduce the Climate Action Rebate Act, which aims to generate $2.5 trillion in tax revenues over ten years by slapping a fee on oil, natural gas and coal starting in 2020.

Senators Coons and Feinstein, as well as many Americans that believe the Green New Deal is a good idea, seem to be oblivious to the fact that 100 percent of the industries that use deep earth minerals/fuels to "move things and make thousands of products" to support the economies around the world, are increasing their demand and usage each year[97] of those energy sources from deep earth minerals/fuels, not decreasing it.

Prosperity around the world from deep earth minerals/fuels is now being weaponized against the West[98] since the oil, natural gas and particularly coal prosperity has led to reduced infant mortality, extended lifespans, and allowed the movement of goods and people anywhere in the world via the diesel engine and jet turbine.

Both have done more for the cause of globalization than anything else; and both get their fuels from oil. Without transportation there is no commerce. Globalized road and air travel dominate most people's lives in industrialized countries and emerging markets.

In addition, the intermittent electricity from wind, solar, or from batteries and storage units made from exotic materials like cobalt and lithium cannot supply the thousands of products from petroleum that are demanded by every infrastructure such as electronics, communications, heating and air conditioning and all the products that are the basis of everyone's standard of living across the globe.

Wind and solar obsessed Germany[99], Australia[100], and Denmark fight it out for the honor of paying the world's highest power prices[101]. America is following, not leading, as they would like you to believe, into known disastrous territory. Just like Germany, Australia, and Denmark before trudging into the green morass, our leaders cannot "see' the direct correlation between energy costs for electricity and fuels, and homelessness and poverty.

Getting-off-fossil fuels would reverse much of the progress society has made over the last few centuries. Until electricity storage technology can support intermittent electricity from wind and solar, the world will continue to have redundant fossil fuel backups[102] for those windless and cloudy days to provide electricity to the world's economies around the clock.

Germany's failed climate goals[103] should be an ominous wake-up call for America and governments everywhere struggling to reach their own targets.

Subsidizing investments in low-power density renewables of wind and solar to obtain intermittent electricity from their huge land mass requirements, are all resulting in higher costs of electricity and fuels to the consumers. In 2016 the U.S. was still subsidizing renewable energy[104] to the tune of nearly $7 Billion. Like in California, Germany and Denmark, the unintended consequences are that the climate goals are further fueling the growth of the homelessness and poverty populations.

Hopefully, BEFORE committing to an all-electric world, we can achieve the technical challenges of discovering a green replacement for the thousands of products based on the derivatives from petroleum. These products are needed by every known earth-based infrastructure. Will society accept the consequences of altering their lifestyles that will result from less services, and more personal input to accommodate losing the advances fossil fuels have afforded them? Maybe the west will commit societal suicide, but it is doubtful China, India, Russia, Iran, or Africa will ever follow that path.

Promises, promises. We're constantly being blown away with the growing capacity of wind farms to provide renewable energy, but they've yet to produce anywhere near their projected capacity. Compounding their lack of production, is the intermittency of what they produce.

AUSTRALIA

- *In Australia, they're losing businesses, jobs, and money - the new definition of madness that's becoming laughable stuff. Australians have become increasingly tired and frustrated with the wind movement. As such, voters went to the poles at their 2019 Federal Election, billed as a referendum on 'Climate Change', and voiced their opinions.*
- *The Aussie Green/Labor Alliance promised an all wind and sun powered future with across the board subsidies for electric vehicles and household battery installations, subsidies for wind turbines, and subsidies for solar panels, and a crippling carbon dioxide gas tax, dressed up as a CO2 emissions reduction target, and an elevated directive for new cars to be electric.*
- *The top-billed reason Green/Labor was supposed to fare so well at the polls, was that Australians are apparently, spending their every waking hour fretting about carbon dioxide gas and believing windmills and solar panels will save the day. Well, apparently not – Green/Labor duly lost the 'unlosable' election.*

- *After convincing themselves not to commit hari-kari, Green/Labor's next move was to bring in Pope Pompous III aka Al Gore (flying first-class, of course, in a jet fueled by a petroleum product) to lecture the Australian voters who had so resoundingly rejected the nonsense of attempting to run on sunshine and breezes.*

- *Gore, the same guy that attended a natural science class at Harvard, and was awarded a D grade for his efforts, stated that Australia can become the renewable energy superpower of the 21st century by exporting renewable electricity. Really? How do you export renewable electricity? We haven't quite figured that one out, yet.*

- *'Pope' Gore is the same guy in 2006 who got everyone's attention by projecting the extinction of polar bears. He's gone silent of late on the polar bear issue simply because their numbers keep rising and may have 'quadrupled' in the last decade. Polar bears aren't going anywhere except to the edge of the ice to feed on the also-not-going-extinct artic seals.*

- *Australian taxpayers forked out more than $320,000 for a Climate Week conference, where the former US vice president "communicated the urgency of the climate crisis". The fear and scare tactics continue: What was formerly "global warming" and then "climate change", is now a "climate crisis". Green/Labor brought in heavy hitter and hypocrite Al Gore to "train" Australia's "climate volunteers" and to "communicate the urgency of the climate crisis".*

- *The global alarm movement has gotten this far, because of the backing of western millennials who've been indoctrinated since their early childhood days in an education system swayed by left leaning administrators. Enjoying living standards unprecedented in world history, they've embraced alarmism as a new secular religion.*

Australia's power plants compare unfavorably with China and India. Australia has twenty-two operating coal-fired power plants, while over half (5,884) of the world's coal power plants[105] (10,210) are in China and India whose populations of mostly poor peoples is roughly 2.7 billion. Together they are in the process of building 634 new ones. They are putting their money and backs into their most abundant source of energy – coal.

- *Hard working Australians that had been providing across the board subsidies for electric vehicles and household battery installations, subsidies for wind turbines, and subsidies for solar panels finally got tired of paying for and showed up at the poles to express their opinions by rejecting the Green/Labor movement.*

California the fifth largest economy in the world[106] by itself is following Germany and Australia by phasing out its nuclear reactors, which have generated continuous uninterrupted zero emission electricity, in favor of intermittent electricity from wind and solar. In 2013 California shut down the continuous nuclear facility of SCE's San Onofre Generating Station which generated 2,200 megawatts of power and will be closing PG&E's Diablo Canyon's 2,160 megawatts of power in 2024.

The overall capacity of all wind turbines installed worldwide by the end of 2018 reached 597 Gigawatt according to preliminary wind power statistics published by the World Wind Energy Association (WWEA)[107]. All wind turbines installed capacity by end of 2018 will cover only 6 percent of the global electricity demand with actual production significantly less than rated capacity, and that's only intermittent electricity.

Judging from the headlines, the world is on track to ratchet up wind and solar electricity and begin the rapid scale-down and ultimate phase-out of fossil fuels. Most energy analysts consider the fossil-fuel phase-out to be a scientific, economic and political fantasy, akin to levitation and time travel, but somehow the movement keeps making news.

The sad part is the ratchetting up of renewables is not the call of energy analyst specialists, it's the call of elected government officials and appointed government personnel supporting their decisions with less than accurate information.

The 2.7 billion residents of China and India are just starting to board the energy train and enjoy the lifestyles that electricity and the products manufactured from those deep earth minerals/fuels, that most of the rest of the world is presently enjoying.

Seems obvious that we cannot rely on wind and solar expected outputs as they can realistically only provide a fraction of their capacity and then, only do it intermittently. Such intermittency requires fossil fuel backup for continuous uninterruptable electricity.

While everyone improves their efficient use of energy and implements conservation to the best of their abilities, the world needs to use the time to diligently develop new technologies to find an energy source or sources that are similar or superior to what the derivatives from petroleum have been providing civilization. Hopefully those new sources will be abundant, and affordable.

> "God, grant me the Serenity to accept the things
> I cannot change; the Courage to change the
> things I can change; and the Wisdom to know the
> difference."

The afore quoted Serenity Prayer came to mind while we were writing this. It seems applicable to the world's citizens who are trying to attain the leadership roles in the save the environment movement before understanding the complexities of the energy picture depicted in the book Energy Made Easy[108]_and the advantages energy as a whole has provided humanity for the last couple of centuries.

Because developed countries have accomplished much in the last few centuries, they have a responsibility, as caretakers for the only planet we live on right now. Understandably, it's hard to imagine the billions of people in underdeveloped countries who have yet to experience anything like the industrial revolution and who are surviving without any of the advantage's fossil fuels are providing to the lifestyles of those in developed countries.

Yes, there are billions of people in undeveloped countries who are currently living in the low economy medieval days that developed countries left behind a century ago after the assimilation in the early 1900s of the automobile and airplane into regular societal structure. They have yet to join the industrial revolution, and without oil and natural gas, they may never get that opportunity.

It's almost impossible to understand that almost half the world[109] of over three billion people live on less than $2.50 a day. Even harder to grasp is that at least 80 percent of humanity lives on less than $10 a day. Today, across southern Asia, portions of Europe and parts of Africa and Australia, there are families attempting to live on virtually nothing. As hard as it is to believe it is a truism.

Can anyone comprehend that the homeless in America may be living a better life than 80 percent of humanity?

Imagine families living in dirt huts with no access to emergency medical care because there is no EMC. Their daily lives are bleak and hopeless. They watch their children, friends and relatives suffer and die early deaths from diseases/conditions that are curable[110] using medicines and treatments brought about by developments using petroleum product derivatives.

With fossil fuels, for the few of us in the developed countries we can now survive in environments all over the world, even harsh ones like Antarctica. Every year, we fell forests and destroy other natural areas, driving species into smaller areas or into endangerment and some even to extinction, because of our need to build more housing to contain our growing population.

Today, the current world population of 7.7 billion[111] is projected to reach 9.8 billion in 2050[112], and 11.2 billion in 2100. How many more trees will fall as unnatural selection takes its toll on the planet? Currently, underdeveloped countries, mostly from energy starved countries, are experiencing eleven million child deaths every year[113], and mainly from preventable causes.

The politicians' silence is deafening about deaths in poor countries that are mainly from preventable causes of diarrhea, malaria, neonatal infection, pneumonia, preterm delivery, or lack of oxygen at birth.

Imagine the future atrocities to humanity for those trying to live in abject poverty if we deny the growing poor the benefits of medicines, heating and countless other developments made possible by fossil fuels, to ever achieve the lifestyle benefits afforded the few in developed countries from all those products we get from fossil fuels.

The Earth has been around 4.5 billion years. While our ancestors have been around for about six million years, the modern form of humans only evolved about 200,000 years ago. Civilization as we know it is only about 6,000 years old, and industrialization started in earnest only in the 1800's, just a few hundred years ago.

For nomadic tribes that ruled over thousands of years, their governmental powers were driven by horses, mules, and camels from the animal kingdom - true medieval economies.

From those horse and buggy days a few centuries ago, those personal and commercial vehicles that did not exist before 1900 are currently estimated at 1.2 billion vehicles on the world's roads[114] with projections of 2 billion by 2035. By some estimates, the total number of vehicles worldwide could double to 2.5 billion by 2050.

Another thing we take for granted is air flight. Our hat's off to Wilbur and his not so congenial brother Orville. Imagine not being able to fly anywhere in the world today? The airlines that did not exist before 1900, transported more than 4.1 billion passengers in 2017 around the world and projections are 7.8 billion airline passengers by 2036[115].

In just the last few centuries every developed nation now has a military that consists of aircraft carriers, battleships, destroyers, submarines, planes, tanks and armor, trucks, troop carriers, weaponry and troops with support structures that need constant transporting around the globe, as well as a multitude of infrastructures and products that provide for a comfortable lifestyle in their homelands.

Developed countries that are wealthier and healthier than underdeveloped countries have become dependent on the more than six thousand products that are manufactured from petroleum[116] and that includes fuel oils for heating and electricity generation, asphalt and road oil, fertilizers that help agriculture feed billions, and feedstocks for making the chemicals, plastics, and synthetic materials that are in nearly everything we use today.

Interestingly, the primary economic reasons that oil refineries even exist are NOT to manufacture the aviation, diesel, and gasoline fuels for today's military and transportation industries, but to meet the demands of society for all those products from the petroleum derivatives that support current lifestyles.

From a forty-two-gallon barrel of oil only about half is for fuels[117] while the rest is used to manufacture petroleum product derivatives that are part of our daily lifestyles. Those billions in underdeveloped countries may not need transportation fuels, but they do need the other half of the barrel of oil for the thousands of products that have enhanced the lifestyle of those in developed countries.

As headstrong as the leaders of the new environmental movements are, they are equally ignorant of what "energy" means and the real reason fossil fuels are integral to the success of developed nations and the necessity of those fuels being made available to up and coming nations who want to enjoy the fruits and comforts of modern society.

Looking back, the climate alarmist's movement started with Al Gore's 2007 movie when he proclaimed the eminent extinction of the polar bears due to global warming. Since the population of polar bears has blossomed over the last decade, we've yet to hear another word from Al Gore on that subject.

Even former President Barack Obama got into the act when he stated that all the glaciers at Glacier National Park would disappear by 2020 due to climate change. As it turns out, higher-than-average snowfall in recent years[118] upended computer model projections from the early 2000's and in 2019 The National Park Service (NPS) quietly removed a visitor center sign saying the glaciers at Glacier National Park would disappear by 2020 due to climate change[119].

The doomsday forecasters are now grasping at new names to rebrand the movement. What was once global warming, is now climate change, climate disaster, global meltdown, climate collapse, scorched earth, climate emergency, and the latest movement, "we don't have time". Like Gore's initial predictions, all the tweets lack the basis for their dismal projections.

The grandparents of millennials may remember from the late 1950's this best-known quote "Just the facts, ma'am." from Sgt. Joe Friday with the TV series Dragnet. A few decades later there was Clara Peller who was a manicurist and American character actress who, at the age of eighty-one, starred in the 1984 "Where's the beef?"[120]. advertising campaign for the Wendy's fast food restaurant chain.

The short emotional tweets from Alexandria Ocasio-Cortez (AOC) with 5.9 million followers, Greta Thunberg with 3 million followers, Al Gore's 3.1 million followers, Tom Steyer's 250 thousand followers, and Jane Fonda's 500 thousand followers, all bumble about the doomsday that's coming.

The tweets are void of any "beef or facts" as to what's going to cause this forthcoming demise. They tweet rhetorical questions and emotional statements, and the millions of followers being brainwashed with scaremongering dogma slurp it up, as environmentalism has become the new religion.

The alarmism over global warming, climate change, etc., is at the forefront of these tweeted fear tactics, but when such alarmist conclusions are openly rebutted, the rebutters are being besieged with oratory that 97 percent of "all" scientists, and even the non-scientific community of 175+ organizations active on climate change[121] believe mankind has played a role in changing the earth's climate.

We have two problems with that 97 percent claim, 1) common sense tells us that no large group of people on our planet could ever reach 97 percent agreement on anything, even the world being round, and 2) shockingly, none of the scientists of the 97 seem to have a name, it's just a holistic group of no-names!

It seems that none of these "97" are able to "talk" specifically about selective microscopic sound bites from vast data that are the supposedly the basis of these dire warnings about time running out and the idea of a twelve year deadline for the annihilation of life as we currently know it.

Obviously, those poor in underdeveloped countries cannot afford to subsidize themselves out of a paper bag and continue to use what's readily available - coal.

It must be the other 20 percent of the population, or about 1.6 billion, in developed countries that are the targets of these climate alarmists rebranding efforts. The tweeters are promoting a global response to the threat of climate change, sustainable development, and efforts to eradicate poverty.

Yet, it's that same 20 percent that have come out of poverty in the last one hundred years as a result of what those deep earth minerals and fuels have provided society, enhancing their lives and improving their standard of living. Basically, the same fossil fuels that are being deprived from the other 80 percent that now live in abject poverty with no hope of reaping the benefits of what prosperous societies are enjoying.

The folks in prosperous societies[122] that have embraced and increased their production of fossil energy have been amply rewarded with greater economic development and growth, and a healthier society. Virtually all diseases are now under control with medications and medical equipment that was not available in the 1800's, before fossil fuels starting to run everyone's lives.

Today, we can live in any weather condition. We also have military equipment consisting of aircraft carriers, battleships, destroyers, submarines, planes, tanks and armor, trucks, troop carriers, weaponry, along with airlines, merchant ships, cruise ships, truck and cars all over the world that dominate the lifestyles of prosperous societies.

The fossil fuel industry would not be needed except to meet the demands of the current users in those prosperous societies. Our belief is that those users are less inclined to go back to living in medieval times without all the amenities that the thousands of products and the various fuels that the fossil fuel industry have been able to fulfill in their daily lives.

We presume the alarmists that constantly refuse to surface from behind their tweet machines to debate is because they have no case to debate the facts that they are using to justify their growing alarmist vocabulary. Unless there's a face to face debate with the supposedly deniers, that have more data than words, we'll never hear both sides of the climate discussions.

It's definitely time for the alarmists to show us "where's the beef" behind their tweets and marches, so the public can decide for themselves to consider the data from each side or just continue to accept the barrage of tweeted words of impending climate disasters that will end life as we know it.

From the extensive data available on temperatures, weather, sea levels, emissions, etc. that several scientists have shared, we don't see the cause for such a dismal outlook for the earth and its civilization. We suspect that classifies us as "deniers". We're willing to join the doomsday parade, but only if the tweeters would come out from behind their tweet machines and "show their cards". Looking forward to face-to-face discussions.

The reality is that there's nothing political about energy. It's about economics. All the people in the electricity industry have been working their whole lives to provide economically affordable electricity to consumers and businesses; and to be arguing about whether we should have fifty percent, or one hundred percent renewables misses the point. Both major parties know we need natural gas, nuclear and hydro for continuously uninterrupted electricity.

CHAPTER TWO

THE PRE-THOMAS EDISON ERA
The World before the light bulb

By Todd Royal

PEOPLE BEFORE FOSSIL FUELS & NUCLEAR GENERATED ELECTRICITY

Before fossil fuels, and the thousands of products made from petroleum derivatives, and electricity that followed, which is dependent on petroleum to produce electricity the world was a zero-sum hellhole that was a war against one another scrounging for food, water, and shelter. Animals and humans had the same fate when cooperation, and self-restraint were never considered in daily existence.

England's industrial revolution began in the late 1700s and the American colonies – pre-United States' existence – were approximately 1776. This watershed epoch in human history was the beginning of global misery finally ending.

To begin with, people were beginning to consider economic freedom, and the organization of markets for the first time when Adam Smith published *An Inquiry into the Nature and Causes of the Wealth of Nations* in 1776.[123] Smith's book was his economic magnum opus, and the first description of what builds individual, and national wealth. It is a must read for classical economics, and economists.

His treatment of human, and economic productivity points towards the need to electrify society, that fossil fuels and nuclear provide. What the Industrial Revolution revealed was how factories needed the products from petroleum derivatives, electrification for society to progress forward, and the inadequacy of the sun, wind, water, animals such as oxen, and slave labor to meet free market demands of society.

For most of human history, economies were arranged into mercantilist arrangements. Mercantilism was dominant before fossil fuels made electricity possible in Europe, and:

- "Is a national economic policy (mercantilism) designed to maximize exports and minimize imports of a nation; and promotes government regulation of a nation's economy for the purpose of augmenting state power. High tariffs, especially on manufactured goods, were an almost universal feature of mercantilist policy. Historically, such policies frequently led to war and motivated colonial expansion."[124]

Electricity was never a part of these mercantilist, war-inducing economies. What these pre-electrical societies produced was famines that ravaged the world once a generation.[125] Plagues were also the norm. Life in the late 1700s and early 1800s was usually hopeless when:

- "In the 1780s, Four out of every five French families devoted 90 percent of their incomes to buying bread. Life expectancy in 1795 in France was 27.3 years for women and 23.4 for men. In 1800, in the whole of Germany fewer than a thousand people had incomes as high as $1,000."[126]

    (Note: Inflation would bring that $1,000 amount to $20,413 in 2020 dollars.[127])

As appalling as conditions were in Europe, Asia had unspeakable conditions, and many would argue large parts of Asia, India, and China today are finally leaving conditions that still find hundreds of millions living in mud huts, drinking polluted water, and burning cow dung and scrounged wood for their fuel and energy. Currently, over two billion people globally do not have reliable electricity, and 600 million of those live in Africa.[128]

Shockingly, of the almost eight billion people living on this planet, we know that 80 percent of them, or more than six billion, are living on less than ten dollars a day.[129] Those underdeveloped locations in the world, mostly from oil and gas starved countries, are experiencing eleven million child deaths every year.[130]

Now imagine that without fossil fuels paving the way for electricity, and the products for modern medical practices were unknown. Most of the planet was unmapped. Plumbing, sewer systems, and modern sanitation facilities that need products from fossil fuels and electricity to run had not yet been invented.

Potable water was rare. Most humans were so ignorant, and uneducated, they had no idea filthy water spread diseases leading to death. Large parts of India, Africa, Asia, and South America still suffer this fate.[131] There wasn't railroads, cars, long-haul trucking, merchant ships, or airlines. Vast irrigation projects, or mass public education did not exist. No sort of standardized units of exchange, newspapers, magazines, libraries, bridges, roads, aqueducts, and uncorrupted legal systems not ruled by Kings, or Monarchs along with medical advances, or anti-corruption campaigns were all foreign concepts.[132]

During this pre-product from fossil fuels, and pre-electrical time, most nations, and city-states were authoritarian, and it is no coincidence where socialism, communism, and authoritarianism reigned, electricity was in short supply.[133]

Before fossil fuels led to electricity, the only form of self-government, or what we would refer to, as democracy was only found in Great Britain, and the fledgling United States. This was a torpid world where women, children, and anything other than Caucasian men had no voice in the say of their lives, or anything resembling representative democracy.

In the 1800s very few private businesses, or corporations existed, since farm-life is how people lived and died. Most people never traveled 100-200 miles from where they were born. Life expectancy throughout Europe hovered between twenty and thirty years of age. The ability to have free speech, freedom of religion, or the choice to be atheist, agnostic, or simply not care about religion was non-existent. An absolute ruler, some form of monarchy, or Kings who were thought of as deities was the standard.[134]

Likely the King, or absolute monarchy controlled "all political, economic, and moral-cultural matters; traditional (western) Christianity, and Judaism lived under severe constraints."[135] The enlightenment never happens without the products from fossil fuels and electricity since the computer, Internet, libraries, and universities need electricity for use.

Today's market economies, which have their birth from the industrial revolution, would never exist without electricity. But electricity could not exist without the chemicals, and by-products that get manufactured from crude oil.

The market economy powered by the products from fossil fuels and electricity unleashed wealth, innovation, employment, and freedom unlike any single force in history. All of this is possible with electricity that is powered predominantly by coal, natural gas, and nuclear generated electricity.

Great Britain had its population and real wages increase 1600 percent between 1800-1900.[136] This growth represented the first-time illiteracy could be tackled. Varied diets involving choice foods, beverages other than sour wine or dirty water, and the opportunity to acquire new skills and vocations that didn't involve agriculture all came into existence during this time frame. These gains that began in the mid-1800s in personal, and societal liberty are still around today, and powered by electricity. Electricity and equal rights go together.

Unfortunately, electricity with its dependency on fossil fuels for its parts, and the personal, and global gains it brings are often misunderstood. Pope Pius XI believed the tragedy of the nineteenth century was "the new spirit of capitalism as being materialistic, secular, and dangerous to religion (control) and the rising spirit of individualism."[137]

Pius XI lamented the church lost the working classes, and maybe he was prescient since today the very parts of the world that gained the most from fossil fuels and nuclear, which led to electricity the last 150 years – the U.S., Great Britain, and Europe – are the least religious people on earth.[138] If electricity brought freedom, and individual opportunity, it has also brought prosperity for the working masses to unfortunately leave religion.[139]

What fossil fuels, nuclear, and electricity cannot solve is abortion doing away with able bodies and workers for growing economies.[140] When societies lose people, they hamper productivity; and abortion is still strongly used and supported by U.S. taxpayers. People should understand abortion is more than a religious issue.

Mistakenly believing electricity will solve all of society's energy, and emissions problems takes a major hit when economies such as China, Japan, Russia, the European Union and the U.S. who are all losing people to abortion, have confidence in fossil fuels, nuclear, or renewables producing enough electricity leading to greater economic productivity.[141] Progressive productivity needs able bodies, and abortion has the opposite outcome.

If it is the religious task of Christians, Catholics, Jews, Buddhists, Hindus, Mormons, and Muslims to change the world – the very task they are assigned by their devotees – it's only achieved by the fossil fuel products that make electricity possible from natural gas, coal, oil, petroleum, biomass, hydro power, nuclear, wind turbines, and solar panels. Progress typically is achieved within religious organizations, governments, corporations, educational institutions, research facilities, and the family. With progress comes electricity, and this breaks through the iron shackles of politics, culture, and economics.

Per the advent of Edison's curiosity, and brilliance, man has achieved upward mobility. Robert Nisbet's book *History of the Idea of Progress* gives the awareness that sophistication of religion, and morality has their birth in understanding history, politics and economics, which are broken down into fiscal and monetary policies.[142]

Both fiscal and monetary policies are simply fancy terms for how to break down public and government monies. None of these economic philosophies were discernible before electricity. The tragedy before us are the 2 billion people globally who do not have reliable electricity are still unable to discern basic economics, history, government, and religion.

POLITICS BEFORE FOSSIL FUELS & NUCLEAR LED TO ELECTRICITY

Consider politics, philosophy, and public policy before electricity and the discovery, and global use of fossil fuels identified as (oil, petroleum, natural gas, and coal), and nuclear generation for electricity. The Greeks thousands of years ago under Pericles attempted to have a direct democracy where only men voted, and women weren't viewed much better than slaves or livestock.[143] The power of the state, tyranny against the individual, and economic stagnation was the norm.

This was true of the Roman's, great Chinese dynasties, and European monarchies. To just understand the putrid, deathly existence under the Roman's read the 6-volume set of books by Edward Gibbon titled, *The History of the Decline and Fall of the Roman Empire*.[144] Roman energy usage relied on the wind, sun, water, oxen, and slaves constituting the largest segment of energy-labor under Roman dominion.

The terms political economy, or limitations against "crippling taxation, heavy bureaucracy, and dreary regulations of state and church," went on for thousands of years.[145]

Free markets, incentivized careers that placed equal value on men and women, contracts that were upheld throughout society, politics, "covenants, suffrage, and the separation of powers" were unheard of.[146] Add today's United Nations (UN) Universal Declaration of Human Rights, and all this hope and recognition of individual rights began when fossil fuels, and nuclear were discovered, put into widespread use, and given the ability to transform cities, counties, states, nations, and continents via electricity.[147]

Nothing has revolutionized human life, eliminated poverty, reduced famine, and taken human choice to unfathomable expectations the way fossil fuels, and nuclear have since the beginning of the industrial revolution in the 1850s. Fossil fuels and eventually nuclear took agrarian societies, and allowed them to become industrialized, urban, sanitary, and spurred economic development that didn't exist 150 years ago using the thousands of products made possible by the derivatives from petroleum and electricity.

Only Alex Epstein's book, *The Moral Case for Fossil Fuels* has attempted to equate morality with widespread energy and electrical use.[148] Whereas Vaclav Smil's book, *Energy and Civilization A History* through physics, engineering, energy mathematics, history, and critical reasoning without bias, or a political agenda, painstakingly explains why fossil fuels, and nuclear are superior to any other forms of electricity generation.[149] Mr. Smil is one of Bill Gate's favorite authors, and wholeheartedly endorses Professor Smil's books, and research.

If religion, or simply wanting to be left alone is inherently personal then what about the individual person before fossil fuel, and nuclear was able to deliver reliable electricity? What was their fate before Thomas Edison's invention of electricity, and the light bulb; the answer is horribly bleak if we return to relying on the sun and wind to power our world.

GREENS ARE KILLING ELECTRICITY

The only thing both authors of this book see stopping forward progression, human longevity, and innovation is the obsession, and unrealistic expectations of the green movement to rid the world of fossil fuels. The irony of the green movement is by ridding the world of fossil fuels they would also eliminate all parts needed for wind turbines, solar panels, and nuclear to function.

Understand, the green movement, touting global warming for political gain, blames the natural phenomenon of the Australian bushfires on climate change in early January 2020. When over 200 acts of arson started the calamitous blazes.[150] Australia's indigenous people had a solution for the country's bushfires that have been around for 50,000 years: The Aboriginal burning practices. Aboriginal techniques were based in part on fire prevention: ridding the land of fuel, like debris, scrub, undergrowth and certain grasses.[151]

A larger, tragic example of electricity's promises, and its misuse comes from United States Agency for International Development (US AID: the main arm of the U.S. government's foreign aid programs), led by Commissioner Mark Green. Mr. Green stated in 2019:

- "Electricity enables access to refrigeration, (which is also dependent of fossil fuels for all the parts of refrigeration systems), to store, fish, milk, and vaccines. Electricity brightens the night and helps school children study. Electricity allows businesses to stay open later and make communities safer."[152]

Mr. Green has given an excellent analysis of what electricity does for our modern world, and prosperous societies that cannot function without abundant, reliable, on-demand electricity overwhelmingly powered by fossil fuels, and nuclear.

Expensive, intermittent renewable electricity does not meet the requirements of on-demand electricity to meet society's demand. It is mathematically impossible when it comes to the number of land-grabbing solar panels, and wind turbines to equal fossil fuel and nuclear energy density, the amount of nuclear plants needing to be built, fossil fuels such as natural gas and coal needed for energy to electricity consumption, and the overwhelming number of solar panels and wind turbines needing to be installed on a daily basis.[153]

Developed countries that do not want to return to the 1800s, or the drudgery of human history before fossil fuels were instrumental in introducing the world to electricity. Otherwise, the unpredictable electricity at higher prices that solar panels and wind turbines deliver would never be able to promise on-demand, continuously and interruptible, 24/7/365:

- "Refrigeration, heat, light, factories, businesses, jobs, modern schools and hospitals, better living standards, longer and healthier lives – then takes it all away, for hours, days or weeks at a time."[154]

US AID highlights a green mindset prevailing over sensible, and life-giving environmentalism. Currently, the average Sub-Saharan African (and recall from the introduction over 600 million Africans are without reliable electricity at this time) only has electricity "one hour a day, eight hours a week, 411 hours a year – at unpredictable times, for a few minutes, hours or days at a stretch."[155]

Under Mr. Green's solution of solar panels and wind turbines bundled into farms to deliver electricity to Sub-Saharan

Africa their situation would only improve by "ensuring electricity maybe 25 to 30 percent of the year: seven hours a day, fifty hours a week, 2,628 (110 days) a year, at totally unpredictable times, thanks to wind turbines and solar panels."[156] This would add to Africa already having the worst electrification rates in the world.[157]

The danger of returning to this Pre-Thomas Edison mindset is an obsession with "low-emission economic development," entirely centered on electricity.[158] Emissions are defined as plant-fertilizing carbon dioxide (CO2), a trace gas that makes up 0.04 percent or 400 ppm of the earth's atmosphere. Life on earth is possible, because of CO2.

USAID isn't the only global agency to center their entire mission on climate change – but the largest economy on earth (the United States) having its development monies centered on climate change – instead of working towards delivering electricity to Africa, or the other two billion people on earth without electricity, and the almost six billion on earth that live on less than ten dollars a day doesn't bode well for eliminating electricity-poverty in the near future.[159]

More troublesome, and why the environmental movement is so damaging to humanity is their belief in clean, green, sustainable renewable electricity when solar panels and wind turbines are terrible for the environment and human rights.[160] Horribly, wind, solar, and energy battery systems require "200 times more raw materials per megawatt (of electricity produced) than fossil or nuclear energy."[161]

Electricity will be extinct if it needs intermittent renewables, and require an extraction of metals, minerals, and limestone that would devastate large parts of the world. Under current technological constraints, solar panels and wind turbines are unable to deliver electrical generation to the U.S., Africa, or rest of the world without destroying the world's eco systems.

Even if the technology were improved, then 100-kWh energy

battery backup systems would need to be deployed that requires child labor for mining purposes.[162] And needs slave labor in other countries for the deep, rare earth minerals required for them to properly function.[163]

USAID, the UN, the European Union, and Green New Deal proponents are dangerously naïve at best. They're avoiding the realities of the eleven million children dying each year in under-developed countries.[164] Their good intentions are killing people, and removing fulfillment, and happiness from their lives that the products from petroleum derivatives and electricity provides.

Solar cannot provide electricity at night, because the sun isn't shining. At its best energy density output, solar only powers society maybe 30 percent of the time year-round."[165] Whereas a 600-MW coal-fired power plant generates roughly 600 MW of continuous and interruptible electricity.

This amount of solar can only generate 200 MW. Wind is worse. Sustained winds over thirty mph are needed to generate at full capacity, which rarely takes place globally, or in Africa. At the time of writing this (January 2020) Africa had "almost 2,500 MW of highly intermittent wind farms."[166]

This is beyond unfair for pampered Americans, and Westerners who have all the energy we need, and not allow Africans the ability to leave eco-imperialism, and colonial type of behavior. A western country under the NATO-security um-brella, Denmark, and it's five million people can waste billions on wind turbines electrifying their country causing the highest electricity prices in Europe; but over one billion Africans who are on their own don't have that luxury.[167] Billions of aid dollars are wasted that is anti-people, anti-development, anti-fossil fuel, and anti-nuclear, while cloaked in racism bordering on eugenics.

For what – climate change – if we want to save the environ-ment then give Africans, Chinese, all of Asia, Indians, and oth-ers without electricity coal-fired, natural gas-fired, and nuclear power plants. Lay the foundation for prosperity that historically leads to clean landscapes, fresher air, and an all-around better earth for all.

But if climate change is based on the science then we have a bigger problem since it was discovered, and reported on the British Broadcasting Channel (BBC) the major issues science is having in the western world with flawed studies, and "growing alarm (within the global scientific community) that results cannot be replicated."[168]

Poverty, disease, and misery were the world's standard before fossil fuels led to Edison's discovery. Are we shattering reliable electricity's existence to make privileged, white westerners happy about themselves sipping cocktails in Malibu, California, eating crepes in Paris, France, and hiking on glamorous African safaris while the black people carry their back packs for them?[169] The answer seems to be yes. This type of eco-colonialism deeply bothers both authors of this book.

SNAPSHOT OF GEOPOLITICS & ELECTRICITY

In January 2020 Russian oil production hit post-Soviet Union records, defied OPEC, and the OPEC+ supply cut deal, when the Russians under Vladimir Putin's leadership ramped up exploration & production (E&P) to hit new all-time highs.[170] Output exceeded agreed upon production limits nine out of twelve months in 2019 while OPEC's production declined in December 2019.[171]

Geopolitics took a positive turn at the same time during Russia's duplicitous behavior when Greece, Israel, and Cyprus agreed, and signed a deal to build a 1,180-mile subsea pipeline that will carry and move natural gas from the Eastern Mediterranean to Europe.[172]

Final investment decisions will be completed by 2022 on the EastMed gas pipeline, and the goal for completion is 2025. This project is opposed by Turkey, but celebrated by the U.S., European Union (EU), and NATO, as a counterbalance to Russian natural gas from the Nord Stream 2 pipeline flowing from Russia to Germany, and other European countries.[173]

But Russian geopolitical intrigue continues with the opening of the TurkStream pipeline in early January 2020, which runs from Russia to the western edge of Turkey via the Black Sea.[174] This pipeline carries natural gas to Western Europe, and is a long distance connection bringing Russian natural gas to China.

What we're reminded of is not geopolitical intrigue, or the movement of nations, countries, and continents against one another, but the importance of electricity. Without natural gas – which is one of the main energy components that produces electricity – these countries would wither away, and eventually die.

Natural gas, and coal meet the criteria of being abundant, reliable, affordable, scalable, and flexible, all components needed to produce reliable electricity. U.S. greenhouse gas emissions (GHG) fell by two percent in 2018 due to increased use of natural gas over coal.[175] Electrical delivery was rarely interrupted throughout the U.S., and the American economy, stock market, and job growth were all moving upwards in 2018.

What do Turkey, Russia, Israel, Cyprus, Greece, Germany, most of the EU, and the U.S. all have in common? All these countries are nothing without electricity that is powered by natural gas, coal, petroleum, oil, or nuclear. Yes, political intrigue ensues over fossil fuel, and natural gas is a fossil fuel, but it is a reminder that these countries would never function without electricity.

Humanity was in an abysmal state before widespread use of fossil fuels and the products manufactured from petroleum derivatives; and nuclear generated electricity when self-government, free speech, and necessities we take for granted did not exist.

CONCLUSION

If earthly wisdom in the book of *Ecclesiastes* reveals, "there is nothing new under the sun," the existence for humanity for billions of years has been sluggish, listless, and without much hope or feelings. Then fossil fuels were found, the wind, sun, animals such as oxen, and slaves were left behind, and electricity was birthed. With products from the petroleum derivatives and electricity came new diets, better healthcare, and respect for human rights to name only a few benefits.

Add six thousand products that come from a barrel of crude oil that is broken down in petrochemical factories using electricity, and the modern world keeps achieving breakthrough after breakthrough.[176] Eventually fossil fuels and technology companies will work together to produce more oil, petroleum, natural gas, and coal cheaper, more efficiently and cleaner than ever before. It is happening now – this marrying of progressive energy, and artificial intelligence would never take place without reliable and affordable electricity and the products from petroleum.[177]

The skeptic could rebut the positivity of electricity by invoking World War I & II, and the endless wars in the Middle East that have all used electricity to kill millions. The birthplace of civilization (the Middle East) continually wars in tribal conflicts, and religious nonsense, instead of invoking science, reason, and diplomacy over terrorism, squalid continent-conditions, and zero progress exemplified by Iran.[178]

It is indisputable that fossil fuels and the products we get from those petroleum derivatives and nuclear have produced electricity that has taken us from no indoor plumbing, and shortened life spans to man walking in space, and eradicating plagues and famines on a global scale.[179]

Electricity isn't morality, but after the brutal European wars of the 17th century – writers that influenced the American Revolution, and the UN's respect for human rights – Charles Montesquieu, Adam Smith, and James Madison (he of "Madison" Broadway play fame) – searched for collaboration over disputes. Fossil fuels and nuclear energy bringing electricity to life have brought a new vision and order the past 150 years.

After death took most lives throughout mankind's existence before the age of thirty, and women and children regularly died during childbirth, electricity brought life. The world was without dreams, or imagination. The practical superiority petroleum derivative products, and electricity brings to people, cities, counties, states, countries, and continents isn't theoretical, but practical, and instantaneous towards a better life, and world.

CHAPTER THREE

POST 1900'S ERA
After oil began to support Automobiles and Airplanes

By Ronald Stein

INTRODUCTION

Before the inventions of the automobile and the airplane, life was hard and dirty. Bad weather and wretched diseases shortened life longevity. Once petroleum proved to be the energy source that could meet the demands of society for fuels, and the thousands of products made from the derivatives from petroleum and electricity, the world changed dramatically to meet the demands of society, and is today in a constant state of technological changes.

Here's a partial list of what we didn't have pre-1900's:

- NO military equipment: aircraft carriers, battleships, destroyers, submarines, planes, tanks and armor, trucks, troop carriers, weaponry
- NO medications and medical equipment
- NO vaccines
- NO fertilizers to help feed billions.
- NO cell phones, computers, and I Pads
- NO vehicles

- NO airlines that now move four billion people around the world
- NO cruise ships that now move twenty-five million passengers around the world
- NO merchant ships, now in excess of 60,000 that are moving billions of dollars of products monthly throughout the world
- NO tires for vehicles
- NO asphalt for roads
- NO water filtration systems
- NO sanitation systems
- NO space programs

On the surface, the Green New Deal (GND) sounds enlightening. Use a nonexistent super grid of renewable intermittent electricity to replace coal and natural gas power plants so we can all breath air with no emissions from electricity production.

The renewable term in all these cases is not energy in its totality, but just "electricity". Wind and solar farms can only produce electricity, and even that is intermittent, as we need the wind to blow or the sun to shine, or both continually as far north as Oslo and as far south as Christchurch.

With a few facts noted below, you decide, but it appears to be a given that electricity alone to run the world is not going to happen.

Electricity has its limitations, not being able to manufacture the derivatives from petroleum that are the basis of the thousands of products,[180] which are the basis of today's lifestyles, and to energize (no pun intended) the societies around the world to support worldwide transportation infrastructures. What I just described is the basis of international commerce.

Let's take a closer look at a few deficiencies of intermittent renewable electricity from wind turbines and solar panels as it's obvious that any "super grid" of electricity will be unable to support the two prime movers that have done more for the cause of globalization than any other: the diesel engine and the jet turbine, both of which get their fuels from oil.

Since we've come a long way from those horse and buggy days just a few centuries ago before the inventions of the automobile and the airplane, let's take a look at a few facts about the industries that are increasing, NOT decreasing, their needs for deep earth minerals/fuels to "make products and move things" each year. You can then decide if the GND is the magic solution to save the world.

We would not be able to "make products and move things" if not for the thousands of products from petroleum derivatives[181] that get manufactured from crude oil that wind turbines and solar panels cannot manufacture. Economies around the world, and all the infrastructures are increasing their demand and usage each year of those energy sources from deep earth minerals/fuels to make thousands of products, inclusive of but not limited to:

- Asphalt for all the roads
- Tires for most of all vehicles
- Fertilizers to help agriculture feed the world
- Steel for every building in the world
- Wire for the worldwide electrical grid
- Electronics for worldwide communications
- Medical cures for most diseases

We've all read about Greta Thunberg, the Swedish teenager diagnosed with Asperger's high functioning autism, and obsessive-compulsive disorder with a mastery of social media, but no college level education to speak of ferries across the oceans in multi-million-dollar yachts made with the derivatives from petroleum in lieu of flying on airlines powered by aviation fuels.

Greta's elite racing yacht, the Malizia II by Pierre Casiraghi, a member of Monaco's ruling Grimaldi family, and the youngest grandson of Princess Grace Kelly would not be able to sail on a still river if it weren't for a barrel of crude oil. Electricity can possibly power the yacht, but it cannot produce the components it takes to build the yacht under current technological constraints.

Hypocritically the yacht helped protect the Earth from emissions from plane trips by Greta, her father, and a cameraman. Trans-Atlantic flights both ways for the yacht's crew and Greta's handlers did add up though.

If the other 4 billion passengers that currently fly on airlines annually would like to travel like Greta and Pierre, you can have your own Imoca 60 class yacht built for around five million dollars (sails extra).

POST THE 1900'S:

All the components that comprise the following infrastructures of airlines, cruise ships, merchant ships, trucking, vehicles, military, and space travel are dependent on those derivatives from petroleum that all occurred after 1900:

> AIRLINES are accommodating four billion passengers per year[182]:
> - More than 40,000 airports[183] of which more than 1,200 are commercial airports[184].
> - There are about 39,000 planes in the world[185] – including all commercial and military planes.
> - The number of flights performed globally by the airline industry increased steadily since the early 2000's and is expected to reach 40.3 million in 2020[186].
> - Airlines that are consuming more than 225 million gallons of aviation fuels PER DAY[187] to move almost ten million passengers and other things PER DAY, and that is increasing every year.

- Boeing, one of the world's biggest aircraft manufacturers, says that 39,620 new planes will be needed over the next twenty years. This estimate puts the number of aircraft in the world at 63,220 by 2037[188].
- Passenger projections in 2036 are in excess of 7.7 billion[189].

CRUISE LINERS are accommodating twenty-five million passengers per year:

- More than 300 cruise liners[190] are consuming around 80,000 gallons of fuel PER DAY, per liner[191].
- As a side note, billions of vehicle trips to and from airports, hotels, ports, and amusement parks are increasing each year.

MERCHANT SHIPS with 60,000 ships in commercial maritime transport[192] moving products around the world:

- The merchant ships fuel consumption is more than 200 tons of fuel PER DAY, per ship[193] to move merchandise around the world.

TRUCKING with more than fifteen million moving products in America alone[194]:

- Fuel consumption is astoundingly more than 140 million gallons of fuel PER DAY[195].

VEHICLES to move people around the world:

- In 2019 a Los Angeles Times article[196], citing Edmunds data, notes that 325,000 electric and plug-in hybrid vehicles sold in the U.S. in 2019, down from 349,000 from 2018. Those dismal numbers represent two percent of the seventeen million vehicles of all types sold in the United States in 2019. The number of battery-electric models available more than doubled last year, but EV sales didn't budge much. Are EV carmakers driving off a cliff?"

California	46.80%
New York	4.80%
Florida	4.20%
Washington	3.90%
Texas	3.60%
New Jersey	2.80%
Massachusetts	2.70%
Illinois	2.20%
Arizona	2.20%
Colorado	2.10%
Virginia	1.90%
Maryland	1.90%
Pennsylvania	1.80%
Rest of U.S.	19.00%

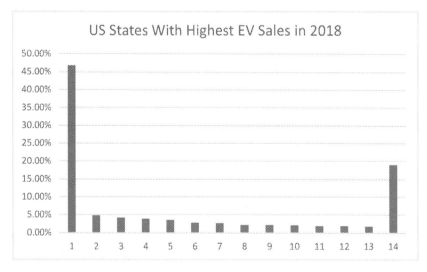

US States With Highest EV Sales in 2018

Table shows that California is the primary buyer of EV's[197] while the rest of America has shown little interest in the incentive and the increasing choices of models.

- Today, there are 1.2 billion vehicles on the world's roads with projections of two billion by 2035[198]. By some estimates, the total number of vehicles worldwide could double to 2.5 billion by 2050.
- Registration of electric vehicles is projected to only be in the single digits, around five to seven percent. If projections come to reality by 2035, five to seven percent of the two billion vehicles would equate to 125 million EV's on the world's roads. The bad news is that would also represent more than 125 BILLION pounds of lithium-ion batteries that will need to be disposed of in the decades ahead.

MILITARY needs around the world are increasing in every country, each year:
- Military needs to move aircraft carriers, battleships, planes, tanks and armor, trucks, troop carriers, weaponry, supplies and anything else needed to assault and occupy nations is increasing each year.

SPACE travel and exploration:
- The world's participation in the space program, is increasing each year. The Saturn V rocket was 363 feet tall and weighed 6.2 million pounds, the weight of about 400 elephants.

Interestingly, the primary economic reasons that oil refineries even exist for societies lifestyles and economies are NOT to manufacture the aviation, diesel, and gasoline fuels for today's military and transportation industries. From one forty-two gallon barrel of oil only about half is for fuels[199]_while the rest is used to manufacture the chemicals derivatives and by-products that are part of our daily lifestyles.

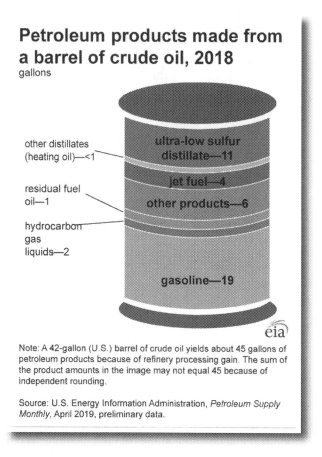

Petroleum products made from a barrel of crude oil, 2018
gallons

other distillates
(heating oil)—<1

residual fuel
oil—1

hydrocarbon
gas
liquids—2

ultra-low sulfur
distillate—11

jet fuel—4

other products—6

gasoline—19

eia

Note: A 42-gallon (U.S.) barrel of crude oil yields about 45 gallons of
petroleum products because of refinery processing gain. The sum of
the product amounts in the image may not equal 45 because of
independent rounding.

Source: U.S. Energy Information Administration, *Petroleum Supply
Monthly*, April 2019, preliminary data.

Those billions in underdeveloped countries may not need
transportation fuels, but they do need the other half of the barrel
of oil for the thousands of products from petroleum[200] that have
enhanced the lifestyle of those in developed countries.

According to the U.S. Energy Information Administration[201]
(EIA), the world energy growth projected through 2040 reflects
the populations of India and China joining an energy society consisting of all those infrastructures dependent on the products and
fuels from crude oil, that continues to decrease its coal usage and
increase its use of electricity from nuclear and from renewables
of wind and solar.

This new energy society of developing countries has yet to

find an energy replacement for petroleum and natural gas. As populations grow, the world's fossil fuel needs continue to increase, albeit slowly, along with their expanding economies and improved lifestyles. Energy.gov reports that Nuclear Power will be needed to meet climate goals[202].

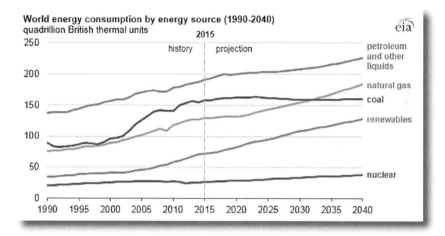

The above graph from the EIA illustrates that electricity alone is unable to support the energy demands of the growing world population and the continuous need for more military, airlines, merchant ships, cruise ships, supertankers, and trucking infrastructures.

Let's also look at the softer side of our lifestyles, other than transportation, as pretty much everything produced by man either includes or requires products made from petroleum derivatives. Even if you go out of your way to avoid petrochemicals that are included in products directly, there's still manufacturing machinery lubricants, plastic packaging, transport fuels, chemical fertilizers, asphalt on roads, and thousands of other things you would never think of, but can hardly live without such as toothpaste and makeup.

The evidence appears to illustrate that electricity cannot

replace the products from the deep earth minerals/fuels that are the basis of today's civilization and lifestyles.

Granted, we need to continue to pursue greater efficiencies and conservation in our daily lives. While we in the developed countries with thriving economies continue to seek out an "alternative energy" that can maintain our lifestyles, the billions of people in undeveloped countries are starting to enhance their lifestyles with the most abundant and cost-effective electricity source available to them today; coal.

Over half (5,884) of the world's coal power plants[203] (10,210) are in China and India whose populations of mostly poor peoples is roughly 2.7 billion. Together they are in the process of building 634 new ones. They are putting their money and backs into their most abundant source of energy – coal.

As those billions in developing countries desire to rise out of poverty and develop modern economies, maybe, by then we'll have a better grasp on a real alternative to those deep earth minerals/fuels that a super grid of renewable intermittent electricity cannot facilitate.

The same U.S politicians that are thrashing on the oil and gas industry, and seeking its demise, are the same ones reaping the benefits of the medications, medical equipment, communication networks, and the thousands of other products from that industry that have contributed to their lifestyles and their ability to live beyond eighty years of age.

They blatantly continue using planes, trains and vehicles to move them and their entourages around the world. Basically, they're enjoying everything they want to deprive from the poorest in developing countries.

The politicians' silence is deafening about deaths in poor countries that are mainly from preventable causes of diarrhea, malaria, neonatal infection, pneumonia, preterm delivery, or lack of oxygen at birth. Those underdeveloped locations in the world, mostly from oil and gas starved countries, are experiencing eleven million child deaths every year[204]. And by the way, adults in those poor countries barely live past forty years of age.

Those children in poor countries still lack purified drinking water, sewage sanitation, adequate nutrition, reliable electricity (or any at all), adequate health care, i.e., the infrastructures and products we take for granted that are all based on deep earth minerals and fuels.

Seems that politicians supporting the demise of the oil and gas industry should speak up and take accountability for supporting the elimination of the industry that could reverse the annual fatality atrocities occurring in those poor countries.

Our U.S. politicians' litigation and regulation grandiloquence seem to be unaware that oil and gas is not just an American business with its 135 refineries in the U.S.[205], but an international industry with more than 800 refineries worldwide[206] that supply oil products and fuels to the world. The constant bickering to attract votes from those even less informed than our elected officials seems to support their desires to hamper or close the American oil and gas industry.

Shockingly, the U.S. could literally turn off the entire country from any source of energy and global emissions would still grow according to U.S. Congressional testimony in 2017[207]. The entire U.S. economy, military and government could disappear, and global pollution, and respiratory illness would still rise. The reason[208] why is "one of the biggest sources of carbon dioxide emissions is developing countries." Think China, India, and Africa.

We're constantly hearing and reading about litigation against U.S. oil firms[209] like Exxon that they knew their industry was a potential harm to the environment. Many of the politicians are targeting lawsuits and further regulations against the oil and gas markets, and a ban on fracking to reduce the crude supplies to those U.S. refineries to hamper their production.

In case you don't remember, we provided at the beginning of this chapter a partial list of what we didn't have pre-1900's.

Both WW I and II were won by the Allies, as they had more oil than the Axis Powers of Germany, Italy, and Japan to operate their military equipment and move troops and supplies around the world.

Post 1900, we now have medications, electronics, cosmetics, plastics, fertilizers, and transportation infrastructures. Looking back just a few short centuries, we've come a long way since the pioneer days. The world had very little commerce before 1900, and without transportation there is no commerce. Road and air travel now dominate most people's lives.

The Energy Information Association[210] (EIA) projects nearly a 50 percent increase in world energy usage by 2050, led by growth in Asia. Even with the rapid growth of electricity generation, renewables—including solar, wind, and hydroelectric power—that is the fastest-growing energy source between 2018 and 2050, the industrial sector continues to account for the largest share of energy consumption of any end-use sector.

The industrial sector includes refining, mining, manufacturing, agriculture, and construction, which is more than half of end-use energy consumption throughout the projection period. The oil and gas industry are projected to remain the main player for world energy, so why are politicians so anxious to tank it?

If it weren't for the demands of societies for the thousands of products made from the derivatives from petroleum and the fuels manufactured from crude oil to meet the demands of the military and the numerous infrastructures to "move things", there would be no need for the oil and gas industry. It's not the supply, but the demand of society that needs to change.

Could it be that our outspoken politicians that are aggressively trying to demise the U.S. oil and gas industry are unbeknownst supporting depriving the poverty-stricken in third world countries from the thousands of products we get from fossil fuels. The same list of products that have virtually conquered all the diseases and allows us to live in any weather conditions and are now the basis of the lifestyles enjoyed only by those in developed countries, while the poorest in the world continue to suffer.

Without the U.S refineries, the USA economy would lose millions of jobs and take the financial hit to have all those products and fuels imported from foreign countries. The world will incur more emissions as most refineries outside the USA, have significantly less environmental controls than here in our homeland. The USA would become a national security risk being dependent on foreign countries for our existence.

With or without the U.S. oil and gas, that's the target of U.S. politician's discourse, the world will continue experiencing enormous fatalities of children worldwide in the energy starved countries.

Do politicians want the oil and gas industry outside America supporting all the transportation needed for international commerce?

We know that you can't ask a logical question to an illogical person and expect a credible answer. It's easy to understand why technically illiterate politicians and political pundits without any science or engineering courses refuse to discuss the basis of their projections. Their jargon sounds good to voters, and they vote the get-off-fossil-crowd into decision-making roles.

Time to stop bickering and focusing on what's important – saving millions of children from preventable deaths.

We never had the oil industry before the 1900's, so why do we believe society can adjust to living in those medieval times with just electricity, but no infrastructures to move things that are the basis of commerce, and no chemicals to make the products that are the basis of our lifestyles?

The Green New Deal (GND) may be in larval form right now, but the fact that it's being seriously discussed in Congress (and around the world) is a quantum leap for politics. If nothing else, the advancing of Ocasio-Cortez and Markey's bill signals that some of our elected officials are on board with inviting

Americans to dream again — to imagine a better future for ourselves, even if the road between now and then hasn't come entirely into focus yet.

Even more important than living without the above-mentioned infrastructures, before the 1900's we had NONE of the six thousand products come from oil and petroleum products[211].

For now, forget about the questions of how to finance the GND's guaranteed jobs for everyone with no infrastructures to work at, and high-quality healthcare for all with no medications or medical equipment.

Just imagine living in those pioneer days with only electricity available and nothing to power since virtually everything we have today is made with the chemicals and by-products manufactured from crude oil.

Turning to the oil industry after 1900 we found that nothing powers economies the way refined oil does; oil can be turned into an array of products from the derivatives from petroleum: cosmetics, athletic equipment, shoelaces, bowling balls, milk jugs, medications and the aviation, diesel and gasoline fuels. The two prime movers that have done more for the cause of globalization are the diesel engine and the jet turbine. Both get their fuels from oil, and without this fuel transportation and commerce return to the pre-Industrial revolution age. In short, oil may be the single most flexible substance ever discovered so why sunset that industry?

Renewables, such as solar, wind, and biofuels, require taxpayer financial subsidies that are derived from the infrastructures supported by fossil fuels, and require countryside-devouring land mass sprawl due to their low-power density to produce significant electricity, i.e., precious land that will be required to feed the billions on this earth.

Commonly proposed alternative energy sources not only can't sustain or improve our relatively high level of prosperity but would lead to economic and social decline in developed countries. Worst of all, the alternatives would leave the poorest in developing countries in continued, unnecessary poverty.

How do we provide subsidies to the renewables industries when at least eighty percent of the eight billion people living on earth survives on less than ten dollars a day? Today, across southern Asia, portions of Europe and parts of Africa and Australia, there are families attempting to live on virtually nothing.

How do we provide healthcare to those children in underdeveloped countries, mostly from energy starved countries, that are experiencing eleven million child deaths every year[212], and mainly from preventable causes[213] when we have no transportation infrastructures to deliver the "medicine man", since there will be no medications and medical equipment without the oil industry.

For those that support sun setting the oil industry that is currently running this world's economy, and embark on those unknown GND roads and negotiate its many turns and obstacles and possibly suffer the consequences of removing an industry before we have an alternative industry to replace it, then keep supporting the GND proponents.

For those that believe we should have an alternative replacement for the oil industry before abandoning the industry responsible for international commerce, then it may be time to change our political leaders.

While developed countries with thriving economies continue to seek out an "alternative energy" that can maintain our economy, the billions of people in undeveloped countries may find difficulty adapting to a world without the oil industry as they are just starting to enhance their lifestyles and commerce.

The poorest countries are the big losers of today's climate activism as they may never get the opportunity to enjoy all the infrastructures and products, we take for granted, that are all based on the derivatives from those deep earth minerals and fuels. With industrial wind and solar proposed for intermittent electricity to replace fossil fuels, third world countries will continue experiencing millions of child deaths every year, and heartbreakingly, mainly from preventable causes.

For those Western politicians, entertainers, and other elites who think climate change is the biggest threat facing mankind, they need to take responsibility and begin to imagine the future atrocities to most of the current world population of 7.7 billion[214] that's projected to reach 9.8 billion in 2050[215], and 11.2 billion in 2100. Most of the poor are trying to live in abject poverty but dying by the millions every year.

For the poorest in the world there are more things that are far more important to survival than climate change. Third world countries are the big losers of today's climate activism. Why? Because they still lack purified drinking water, sewage sanitation, adequate nutrition, reliable electricity (or any at all), adequate health care, i.e., the infrastructures and products we take for granted that are all based on deep earth minerals and fuels.

Yes, there are billions of people in undeveloped countries who are currently living in prehistoric days of existence that developed countries left behind more than a century ago after the assimilation in the early 1900s of the automobile and airplane into regular societal structure. They have yet to join the industrial revolution, and without oil and natural gas, they may never get that opportunity.

Depriving the poorest from all the things we take for granted that "move things and makes products", i.e., the same sources that the world emissions crusaders are trying to replace with renewable electricity. Could it be that climate activists are completely oblivious that wind turbines and solar panels cannot produce any of the six thousand products we get from fossil fuels that are the basis of our lifestyles?

Currently, underdeveloped countries, mostly from energy starved countries, are experiencing eleven million child deaths every year[216], and mainly from preventable causes. Efforts to reduce emissions may do more harm than good to the almost eight billion on this earth, especially to the world's poorest.

Life in third world countries is a clone to life in the Middle Ages, littered with the dusty debris from the detritus of daily life. People and farm animals running everywhere, most with no potable water. Children dressed in filthy clothing. Garbage strewn in the streets, mostly unpaved. These poor people have little hope of ever lifting themselves out of real poverty.

For the minority of folks in the world that occupy the developed countries, they have big problems of their own. Like home Internet quitting for fifteen minutes? Bad mobile signal? Online deliveries arriving late? Can't fast-forward live TV? A closet full of clothes but "nothing to wear"? Sitting near an infant on a flight to Bali? Chipped nail polish?

Can anyone comprehend that the homeless in America may be living a better life than 80 percent of humanity?

We're constantly reading in the paper that Greta Thunberg needs to sail to international climate activism events as she refuses to fly jet planes to her destinations, but what does she suggest for alternative modes of transportation for the 4.1 billion passengers in 2017 that the airlines transported around the world? I wonder if Greta is spending so much time sailing around the world that as a high school dropout statistic, she may become unemployable?

Sure, the politicians and bureaucrats in the developed countries care about climate change—because they expect a piece of the $100 billion-a-year pie of Western "reparations" the Paris agreement promises them. Concurrently, they also express no concern for the world's poorest.

In their minds, climate change/global warming activism is only a "first world" problem.

It's almost impossible to understand that almost half the world[217]— over three billion people — live on less than $2.50 a day. At least 80 percent of all humanity lives on less than ten dollars a day. Today, across southern Asia, portions of Europe and parts of Africa and Australia, there are families attempting to live on virtually nothing. As hard as it is to believe it is a truism.

The economies in developed countries continue to fund climate activism that will keep third world countries in abominable conditions and prevent them from joining the industrial revolution.

We've had more than 100 years to find an alternative or generic to fossil fuels and all the products we get from them, but have only come up with industrial electricity that can be generated intermittently from sunshine and wind, but yet to find a replacement for the source of those thousands of products that are now the standard for our lifestyles.

Over the last 100 years, climate-related deaths in developed countries have decreased by ninety-five percent, mostly from fossil fuels, has lifted more than a billion people out of poverty in just the past twenty-five years." We can thank fossil fuels and capitalism for that and more.

If we continue to deny the growing poor population the benefits of medicines, heating and countless other developments made possible by deep earth minerals and fuels, to ever achieve the lifestyle benefits afforded the climate activists then we need to justify our reasoning for allowing those millions of preventable deaths from occurring every year in third world countries.

The popular climate discussion has the issue backward. It looks at man as a destructive force for climate livability, one who makes the climate dangerous because we use fossil fuels. In fact, the truth is the exact opposite; we don't take a safe climate and make it dangerous; we take a dangerous climate and make it safe. High-energy civilizations, not climate, is the driver of climate livability."

CHAPTER FOUR

UNDERDEVELOPED COUNTRIES FUTURE
Continuation of Deprivation of Products from Crude Oil

By Todd Royal

INTRODUCTION

Over six thousand products come from a forty two gallon barrel of crude oil, and the poorest parts of the world are being deprived of these products that save lives, cities, states, nations, and continents; and ultimately leads to reliable, and affordable electricity.[218] The world is spending valuable resources on wind turbines when every part of the turbine is dependent on oil.[219] The same goes for solar panels.[220] This crushes electrical generation delivery for underdeveloped regions of the globe.

Both sources of electricity generation (solar panels and wind turbines) are dependent on the derivative products that are manufactured from crude oil. Both would not exist without the electricity that overwhelmingly is produced from coal, natural gas, and nuclear energy. The U.S., and world are foolishly beginning to derive most of the new, but intermittent electrical generation from the sun and wind.[221]

Francesco La Camera, Director-General of the International

Renewable Energy Agency believes "investment in renewable energy (wind and sun) needs to double over the next decade in order to hit climate targets."[222] Michael Shellenberger, Time Magazine's Hero of the Environment shows data that everywhere renewables are widely operated they increase emissions?[223]

If we want to lower emissions, and meet the basic requirements of electricity being abundant, affordable, reliable, scalable, and flexible then only natural gas-fired power plants can accomplish this task. If you add zero-carbon to this list, then only nuclear energy meets all requirements.

It is why the United States was the only industrialized country and economy to meet the Kyoto Protocol environmental standards, since natural gas is extensively used for electrical generation, transportation, shipping, and commerce.

For renewables to work they need to overcome darkness when the sun isn't shining (generally cloudless days are best for solar), or the wind isn't blowing between twenty to thirty miles per hour. Renewable electricity also causes prices for electricity to roughly double in price for ratepayers and end-users.[224]

Environmental degradation is also an important issue that still needs to be solved for when during the value and supply chain process solar panels and wind turbines for electricity uses thousands of tons of steel, concrete, fossil fuels, petroleum derivatives, and exotic minerals during the mining process.[225] None of these factors assist delivering, life-giving electricity to poor countries.

Al Gore's chief scientific advisor, Jim Hansen, shockingly said this about renewables and fossil fuels:

> "Suggesting renewables will let us phase rapidly off fossil fuels in the United States, China, India or the world as a whole is almost the equivalent of believing in the Easter Bunny and Tooth Fairy."[226]

U.S. energy policies are moving quickly to shut down all domestic, coal-fired power plants when they should be in business to export coal to underdeveloped and developing economies such as China, India and Africa desperate for electricity that coal provides.[227]

Ask yourself, would you rather have the U.S. with its stringent environmental protections, as it relates to mining, and exporting coal, or China, India and Africa, who have little to no emission controls in place mining, using, and exporting coal? China, India, and the African continent are using more coal for electrical energy since their population growth, and economies are dramatically growing.[228] China, India and Africa make up approximately 4 out of every 7 people on earth.

Ridding the world of fossil fuels and nuclear by only embracing electricity that is produced by renewables (wind turbines and solar panels) is impossible simply by applying the law of physics.[229] Without fossil fuels, and nuclear electricity the world does not have the power to attempt the conversion to only using electricity for every facet of modern life.

Discounting the fact that electricity cannot make the over six thousand products that come from a barrel of crude oil. What are the other major issues underdeveloped peoples, nations, and continents face when they do not have reliable electricity?

UNDEVELOPED COUNTRIES STARVING FOR ELECTRICITY

The world is increasingly embracing energy technologies that are still unable to perform the requirements that alleviate poverty. This isn't a scourge against renewables, but a reality check to understand that under today's technological constraints only fossil fuels, and nuclear can lift undeveloped countries out of energy poverty. India has lifted over 271 million people out of poverty in a decade, because they are finally receiving reliable electricity, mostly from coal-fired power plants.[230]

Global organizations such as the Union Nations, European Union, and the Association of Southeast Nations (ASEAN); particularly ASEAN, when according to the Financial Times we are in the "Asian Century," should be advocating for energy poverty to be alleviated via fossil fuels, and nuclear.[231]

An all-of-the-above approach for energy means coal, natural gas and nuclear are the dominant electricity sources while renewables have research monies invested into them to figure out a more reliable and energy dense solar panel, a better wind turbine requiring less raw materials, and a battery storage systems that can store electricity for days, weeks, or months at a time.[232]

Instead, trillions are being allocated, invested, and spent on undefined climate policies - when the very prediction these trillion-dollar-policies are based on – have never come true.[233] The United Kingdom is an example of a western-aligned country believing it can be carbon-neutral, and demonizing energy companies over climate change.

The chief energy regulator in the U.K. said the oil and gas industry's "social license to operate" is under threat; without ever explaining how to keep their electrical grid from blackouts, or how to replace the products that everyday life, and renewables need to operate?[234] How does this regulators threatening actions bring people out of agrian-based poverty, and deliver electricity?

What coal, natural gas, and nuclear provide that renewables can't is reliable, continuous, uninterruptible electricity. Undeveloped countries are begging for fuels, products from petroleum, and electricity. Africa is a continent with over a billion people where women have too many children per economic opportunities, and the main source of energy is wood and cow dung that results in elevated emissions and respiratory illnesses.

They need electricity, and the derivatives from petroleum to move people, heat materials like steel and silicon used in modern infrastructures, farm machinery to grow food more efficiently using less land and have safe water and sewage treatment plants. These are basic human rights that electricity provides.

All of the components of six thousand daily products, oil, petroleum, transportation fuels (trucking, shipping, airlines), natural gas, coal, nuclear, biomass, hydro power, wind turbines and solar panels fall under the laws of gravity, physics, natural boundaries, friction, inertia, mass and thermodynamics that fossil fuels, and nuclear provide. Without crude oil you do not have renewable electricity.

In this world the more western governments, and affiliated organizations that count the U.S., Europe, some parts of South America and Mexico along with U.S. Asian allies (Japan, South Korea, Taiwan, Philippines) move in the electricity direction from renewables then undeveloped nations will suffer. Increasingly, these countries will then look to Russia, China, Iran, and North Korea for any source of life-giving support.[235] The west fought a Cold War for over fifty years so that would not happen.

Looking to China for support is arguably the most dangerous prospect when approximately a million Chinese Uighur Muslims, in China's Xinjiang Province, have been detained, and sent to re-education camps.[236] The United Nations, and global media outlets are ignoring this "arbitrary detainment" of a predominantly Turkic-speaking group of people.[237] Allowing underdeveloped countries desperate to buy energy from China allows them to continue funding state-sponsored repression.[238]

Underdeveloped countries do not need empty Twitter-hashtag-slogans, or clever software, but fossil fuels, and nuclear energy leading to reliable electricity. Our world now has more cars, planes, long-haul trucks, cargo ships, and factories that are increasing in speed, and consumption of fuel allowing our interconnected, global society to expand.

China, India and Africa are leading this charge, and other undeveloped countries want the same opportunities. These three economies are growing faster and using more global energy and electricity than any other country or multinational organization; and is "projected to grow faster than the rest of the world through 2040."[239]

Electricity alone – which is nothing unless it is given products from petroleum derivatives, nuclear, or renewables – cannot support militaries, airlines, merchant ships, cruise ships, medications, medical equipment, electronics, communication equipment, tires, asphalts, and fertilizers. Underdeveloped countries know this, and it is why they will turn to nefarious nation-states like China to fulfill these needs

Modern economies that desperate countries, such as Pakistan and Mexico, where terrorism and narco-trafficking proliferates in many parts of their countries unabated makes them test cases for electrical growth.

This is likely why Pakistan is aligned with China to build coal-fired power plants under the China-Pakistan Economic Corridor (http://cpec.gov.pk) arrangement.[240] These are the dirtiest, emission-spewing type of coal-fired electrical generation plants being built anywhere on our planet.[241] The Pakistani's are desperate for electricity.

Mexico is in a similar, desperate energy, electricity and national security dilemma. The U.S. may have to intervene sometime soon when it was reported in October 2019, "Mexico is in a state of collapse after a drug cartel defeated the Mexican military in battle."[242]

While exploration and production (E&P) of U.S. shale deposits in Texas, New Mexico, Pennsylvania, Oklahoma, Colorado, and North Dakota grew in 2019, the Gulf of Mexico with sovereign drilling rights held by the government of Mexico is the new "oil boom."[243]

Mexico may sign the Paris Climate Accords but drill relentlessly if it protects their sovereign national interests. Yet California, sitting on the largest oil reserves in America, has chosen to ban fracking and remain the only state in the continental United States that imports most of its oil needs from foreign countries.

These transnational, existential problems take a direct hit when oil and natural gas E&P rises leading to greater electricity proliferation. Many parts of Pakistan and Mexico are primal, tribal, and still stuck in the dark ages. They don't have the capacity, or economies to fight the terrorist, or drug gangs, but they can use coal, oil, and natural gas to fight an economics-based, proxy war with minimal troops. This global, geopolitical fight is done with petroleum; even nuclear is too expensive for these nations.[244]

UNWISE CLIMATE POLICIES DESTROY THE POOR

Greta Thunberg is the teenage, European, climate activist who has been treated for mental illness and an eating disorder. Disturbingly, this naïve little girl has been manipulated by adults, and her parents for their own despicable, financial gains.[245]

Now it's been confirmed a "Facebook bug" found out Greta's father, Svante Thunberg, and Indian climate activist Adarsh Prathap who serves as a delegate at the United Nation's Climate Change organization have been writing her climate posts on Facebook.[246] This child needs to be in school, and study reading, writing, literature, mathematics, art, anything other than climate manipulation by evil adults.

She is a pawn that further stifles underdeveloped countries, and continents all for the sake of money, and climate-power. *The New York Times* also added to the untruthful manipulation when they gave space to an opinion peace in mid-January 2020 where the entire article was dedicated to how you could become a disruptive climate activist.[247]

The New York Times article outlines a five-step plan to help with climate-stress, and only give electricity solutions involving unstable wind turbines, and solar panels. Caucasian billionaire, and U.S Democratic Presidential candidate (2020 election) Michael Bloomberg believes he can achieve a "Buck for Boilers" program similar to former U.S. President Barack Obama's "cash for clunkers (older vehicles)" that failed.

Mr. Bloomberg wants to eliminate fossil fuels, and achieve zero-carbon electricity for cooking, and promote heating water by having pollution-free appliances. He also wants zero-carbon furnaces through U.S. and international taxpayer subsides, incentives, and changing energy regulations. Buildings would achieve zero-carbon status by 2025 under his plan. Mr. Bloomberg has not given any specifics for how these energy and electricity policy recommendations would become a reality.[248]

Bloomberg made billions using fossil fuels to power his media empire, but the poorest in the world are not allowed this opportunity for energy and electricity to alleviate their crushing poverty.[249] The Council on Foreign Relations admits the world is "awash in oil," but doesn't advocate for fossil fuel use when the very offices, and computers this think tank works from has their origins in crude oil.[250] Yet they continue giving misguided energy directives and policies recommendation for increased renewable use that crush the poor.

Neither the Council on Foreign Relations, Bloomberg, or the other billionaire-global-warming-advocate running for U.S. President on the Democratic ticket – Tom Steyer – don't have answers for why the Crescent Dune thermal solar plant in central Nevada – went bankrupt in early January 2020.[251] Costing taxpayers millions in wasted monies.

Nor could Steyer, or Bloomberg explain why James Hansen, the scientist who was one the first proponents for anthropogenic (man-caused, because of fossil fuels) global warming said:

"Renewable energy is a grotesque solution for reducing emissions. The reason is that it costs $140 per ton of emissions prevented by using solar energy."[252]

The German government, however, led by Chancellor Angela Merkel is doing more to destroy poor people, and nations than possibly anyone else on earth. Ms. Merkel while noble in her pursuits of a cleaner environment, and stewarding Europe towards positive goals, has pursued her energy, and climate polices with dreadful results.

International consultancy firm, McKinsey & Company did a study in 2019 on Germany's energy policy called "Energiewende," which is the German government attempting to transition from fossil fuels to just clean electricity.[253] McKinsey & Company findings concluded: "Germany's Energiewende, or energy transition to renewables, poses a significant threat to the nation's economy and energy supply."[254] If German continues faltering then it will fall further under Russian domination, and that has severe consequences for European, Central Asia, and Middle Eastern stability.

Leading German newspaper, *Der Spiegel*, calls the energy transition a "disaster," and shows how trying to rid themselves of fossil fuels, and nuclear led to higher use of coal-fired power plants.[255] These unwise decisions that lead to less opportunities for underdeveloped countries is again exemplified by the German government "planning to spend $44 billion over four years to help the country (Germany) cut its carbon dioxide emissions."[256]

Please understand the Germans would only reduce global temperatures 0.00018 Celsius in a hundred years.[257] The Germans should spend $44 billion on meeting their NATO defense requirements, and upgrading their military to protect against global terrorism, Iran, and Russian bullying.[258]

Germany feeds into the European Union's climate plans to alleviate emissions through advocating for eighty percent reductions by 2050.[259] The conservative, per annual average costs are approximately $1.4 trillion without Europeans learning any lessons from the technologically advanced-Germans failure at their "Energiewende."[260]

This trillion-dollar number does not eradicate the over six thousand products that come from petroleum derivatives; negating any emission reductions the Europeans seek to achieve. Why not eradicate tuberculosis in Africa, and developing nations in Asia "saving more than ten million lives?"[261]

This green obsession is a global phenomenon. New Zealand's government has also promised unachievable, net-zero emission targets by 2050 leading to a zero-carbon society. New Zealand's government has not explained how this will take place under current and future technological constraints? To the New Zealand government's credit, they sponsored a report finding, "the cost of meeting this goal would be greater than the entire current national budget, every single year (until 2050)."[262]

Then there is Mexico with extensive deep water oil finds in the Gulf of Mexico have made promises to cut its emissions in half by 2050 at an approximate cost of seven to fifteen percent of GDP.[263] Again, no coherent, emissions-cutting plan in its place for exactly how this will transpire before this announcement was made.

But nothing is more expensive, or a waste of resources than the 2015 Paris Climate Agreement (PCA).[264] The PCA:

> "Aims to wean entire economies off fossil fuels (never addressing the six thousand products dilemma), even though solar and wind remain uncompetitive in many contexts. The agreement will slow economic growth, increase poverty, and exacerbate inequality."[265]

A new study in World Development via Science Direct that was published in October 2019, revealed the PCA would increase global poverty by four percent, and that "stringent (emissions) mitigation plans may slow down poverty reduction in developing countries."[266]

The United Nations Intergovernmental Panel on Climate Change (IPCC) came to the same conclusions when they mapped five, different future scenarios under the PCA. The IPCC concluded, underdeveloped countries, nations, and continents were markedly more advanced in a "fossil-fueled development" world/economy than under a "sustainable" lower-CO2 planet.[267]

The lower-CO2-sustainable world model would average roughly "twenty six million more people per year in poverty until 2050 than the richer, and less unequal fossil-fueled world."[268] Even with rising climate change worries the findings are still consistent the PCA would increase global poverty, and inequality without reducing emissions in a statistically, meaningful way.[269]

The IPCC estimated climate change does very little to alter economies, or income. Under a do-nothing scenario to counter, undefined, climate change, average incomes drop "0.2 to two percent" by the 2070's.[270] However, incomes are estimated to have risen 300-500 percent by that time.

The hysteria continues unabated over climate change when the Organization for Economic Cooperation and Development (OECD) found "about (70 percent) of all development spending goes to climate projects."[271] These "catastrophic visions" will only lead to higher emissions, and increases in respiratory illnesses globally.[272]

Aid monies could be spent more effectively. Using Africa as an example, if tuberculosis, and family planning are continent-wide initiatives then poverty reduction leads to better climate policies for over a billion people.[273] It seems backward that better economies lead to better environmental health, but that is exactly what happens when:

"History has shown conclusively that making people richer and less vulnerable (giving them reliable
electricity) is one of the best ways to strengthen
societies' resilience to challenges such as climate
threats."[274]

Nobody is more vulnerable to natural disasters, or the threat
of climate change than poor, exploited villages, towns, cities,
counties, nations, countries, or continents than undeveloped or
developing peoples.[275]

Cutting emissions will not help them, or the developed world
in the process. The west is destroying the poor, and killing human
longevity over emissions, faulty environmental agreements, and
trying to eliminate fossil fuels and nuclear generated electricity.

EUROPEAN & INTERNATIONAL SECURITY ISSUES

Green, socially, and environmentally conscious parties are gain-
ing in popularity, status, and parliamentary power throughout Europe. But Europe remains "the world's largest importer
of oil and natural gas."[276] This has devastating consequences
for developing countries having reliable electricity, European,
continent-wide security that involves NATO, and overall peace-
ful development and tranquility since the end of World War II.

With oil output in maturing wells, and oil and natural gas deposits off the coast of Norway declining, European countries are
acquiring oil, coal, and natural gas anywhere they are available.
As the previous section noted, environmentalists, and their allied
climate change doomsayers are killing global, and now European
exploration and production (E&P) for oil, and natural gas. This
shouldn't be the case for Europe.

The possibilities exist there are more shale oil and natural gas
reserves in Europe than the United States, which is "currently
the world's largest producer of both oil and natural gas."[277]

Horizontal fracturing (fracking) in most European countries is illegal or faces daunting legal hurdles, and cultural protests.

As Europeans gave into militant-authoritarianism in the 1930s leading to global war, it presently means the Europeans are giving into authoritarians by needing Russia, the Middle East, Africa, and other despots in the world to supply their energy, and electricity needs.

Particularly, the Russians, and the geopolitical, national security problems the Europeans have brought on themselves with the Russian-owned, Nord Stream 1 and 2 pipelines.[278] Vladimir Putin's weaponization of energy has the Europeans in a stranglehold when over 30 percent of all natural gas supplies come from Putin's Russia.[279]

Couple this with the drop off in oil and natural gas supplies from Iran, and Venezuela, and Europe is reluctant to abide by their NATO obligations, or continent-wide security issues. Even wealthy Europeans panic when electrical grids shut down over relying on intermittent renewables over fossil fuels or nuclear, or a lack of natural gas in frigid winter months.

No one is immune to the disaster that occurs from not having reliable electricity, and abundant energy supplies. Now imagine the billions of people globally without reliable electricity, or none?[280]

The Europeans cannot guarantee energy supplies when the majority of European-NATO members do not meet their basic, defense spending obligations.[281] If the Strait of Hormuz is closed down by the Iranians over the American-Iranian standoff, the Europeans have no choice, but make sure the U.S. Navy guarantees their maritime electricity security. This is one of the main reasons the Europeans choose the Iranians, and the Iranian nuclear agreement over traditional western allies, and NATO.

European aggrandizement of fossil fuels and nuclear is one of the worse situations in our modern, geopolitical world. Europe will not develop its own oil and natural gas reserves, or shale deposits, thus allowing manipulation by importing energy from human rights abusers, and hostile actors on the world stage.

Europe will continue angrily rebuking Israel (the only democracy in the Middle East and respecter of gay and women's rights) in world media outlets, and the UN, because they need Middle Eastern and North African oil and natural gas for energy and electricity stability. They are also uneasy about U.S. President Donald Trump and his "Make American Great Again," vanguard slogan.

Under Trump, the U.S. no longer wants to be the world's policeman, and no longer needs oil and natural gas from Middle Eastern authoritarians. Why should America ensure commercial shipping lanes, and safe flights for the Europeans to bash them continually, and then go use Russian, Chinese, and Iranian oil, and natural gas whenever it suits their energy, and electricity needs? Go examine Trump's America First Energy platform for verification.

What we have is America letting Europe alleviate the messes of their own making. Trump seems to indicate Europe won't develop their own energy and electricity sources, then let them deal with nefarious Russia, the Iranian mullahs, and the duplicitous Chinese.

European irony abounds when they will ask to be defended by American-led NATO to keep the fossil fuels flowing while bad-mouthing Trump, and America; but Trump, the U.S. Congress, and American people have made it clear this is no longer in American interests to engage in European rescue missions. It is anyone's guess whether the Americans under a Republican, or Democratic administration will come to their rescue over energy, and electricity?

The irony is the Europeans would never speak out, or denounce Christian genocide taking place in the Middle East – so as not to offend their oil-exporting autocracies they are propping up.[282] Europe says one thing about human rights, but does another when it comes to energy, and electricity.

Europeans give grand lectures about greenhouse gases but is the world's largest importer of oil and natural gas from unsavory characters and regimes. Their denunciations of fossil fuels and nuclear on the world stage is a hypocritical charade that is an international security risk.

U.S. environmentalists, the green-aligned, Democratic Party, and American multi-national companies are just as guilty of national security instability, as the Europeans. In the January 2020 State of the State speech by the Colorado governor, environmental activists interrupted the Governor's speech by calling for a "Fracking Ban."[283] Colorado already has one of the toughest oil and gas regulations in place, but climate change extremists want more.

No other industry produces more jobs within the value and supply chain than oil and natural gas production. This means more taxes for infrastructure, education, and public safety. These activists do not have answers for where these tax revenues would come from if they were successful, and shut down oil and natural gas E&P.

Big business is now pushing the electricity-cure-all to easily grow their businesses off the backs of taxpayers, and the uninformed public. This is also great for their stock prices, balance sheets and income statements by having access to tax credits and subsidized taxpayer electricity mandates by supposedly being green.

BlackRock, the world's largest money manager, is overhauling its investment strategy "to make sustainability the 'new standard of investing,'" according to its CEO Larry Fink.[284] Mr. Fink wants to "fundamentally reshape finance," since climate change has "become a defining factor in the companies' long-term prospects."[285]

To Mr. Fink's credit he did acknowledge that after rolling out new products that focus on sustainable investing (whatever that means since Mr. Fink or BlackRock didn't give specific details) that:

> "Under any scenario, the energy transition will still take decades. Despite recent rapid advances, the technology does not yet exist to cost-effectively replace many of today's essential uses of hydrocarbons. We need to be mindful of the economic, scientific, social and politics realities of the energy transition."

Likely, BlackRock will continue investing heavily in Chinese coal-fired power plants, and India's nuclear reactors. Add U.S. shale play investments, and BlackRock never having an answer for how you rid themselves of the six thousand products that come from a barrel of crude oil, and Mr. Fink's environmental activism looks self-serving, and rent seeking for U.S. tax dollars.

Other misguided, even dangerous U.S. companies include Microsoft, and Starbucks. Microsoft President, Brad Smith, at the Davos, Switzerland economic conference in January 2020 "pledged to eradicate its carbon footprint."[286] Mr. Smith didn't define how he would accomplish this task when it is well-known Microsoft's cloud computing services are some of the largest users of electricity in the world, and need continuous, uninterruptable electricity, not unreliable, intermittent electricity from wind and solar.

Starbucks, which gathers coffee beans from unsustainable poor nations globally that are not concerned with Starbucks' definition of climate change, or advocacy of renewables for electricity, "has announced multi-decade aspiration to become resource positive."[287]

Does this mean some of the poorest countries in the world where Starbucks acquires its coffee beans will only use wind turbines and solar panels for electricity? Will Starbucks give up all six thousand products from crude oil? Will Starbucks limit water intake for migrant workers picking their beans? What about housing – will Starbucks build carbon-neutral housing that doesn't use products from crude oil?

If so, will Starbucks invest in carbon neutral concrete, steel, and rebar; or will they only use straw housing that is grown from well water? Starbucks has not defined how this aspiration will be accomplished.

More concerning and frightening for global security, and the post-World War II, U.S.-led, liberal order is the leading environmental organization in the world – The Sierra Club – wholeheartedly promoting "wind and solar push threatening US power supply."[288]

In a December 27, 2019 titled, "Federal fossil-fuel subsidies hurt renewables, drive up our electric bills," the Sierra Club's New Jersey President, Jeff Tittel accuses the Federal Energy Regulatory Commission (FERC) of a "shameful giveaway" to oil, coal, and gas interests over promoting solar and wind farms energy to electricity.[289] Current U.S. renewable energy to electricity usage only constitutes "8 percent contribution by wind and solar."[290]

Electrical grids need reliable electricity, and when grids have unstable electricity they black out. Grid collapse is guaranteed under "100 percent renewables by 2035."[291] If the U.S. didn't have electricity, we could witness a global catastrophe on the scale of World War II. Whether we like it or not, the United States military, its diplomatic soft powers, and economic might have kept the world from global wars for over seventy-five years.

If the United States, and Germany cannot figure out renewables, and how to contain electrical grid blackouts then what hope do the world's most vulnerable peoples, nations, and continents have for their countries and future? A bleak one if they are forced to adhere to the western energy model based on climate change, solar panels, and wind turbines.

Renewables, and climate change policies are not the answer. Fossil fuels, and nuclear are the moral choice, and best-case scenario for them to secure abundant, affordable, reliable, scalable, and flexible electricity. Otherwise, governments, business leaders, and global leaders could find a wave of "populism, nationalism, and protectionism," sweep over them; and this movement doesn't care about global emissions, or environmental treaties.[292]

CONCLUSION

If you want solar panels and wind turbines for emission reductions and eradicating global poverty simultaneously then invest in the renewable sector. Attempt to drive the costs of solar, and wind electricity down to the level of fossil fuels, and nuclear energy to electrical generation. Do this without any taxpayer subsidies, or mandates forcing electrical grid operators to purchase electricity from the wind and the sun. Then find a way to do away with every part in the value, and supply chain for solar panels, and wind turbines no longer having their beginnings from a barrel of crude oil.

Otherwise, the west led by the United States, European Union, and the United Nations are killing the most vulnerable people in the world without benefitting the environment, economic health, human longevity, and possibly leading to longer and more severe wars chasing after scarce resources. A dangerous fantasy is underway that electricity alone can cure all our societal ills.[293]

Give these susceptible and economically defenseless people and continents in locations such as Africa abundant, reliable, scalable, affordable, and flexible electricity; instead of climate-solutions, which cause more problems than they alleviate.[294] Most of all the west should understand some basic truths about the climate and renewables that is dramatically affecting reliable electricity expanding globally.

Al Gore, former UN Secretary General Ban Ki-Moon, Greta Thunberg, and in January 2020, New York Times columnist Paul Krugman, all declare the world will end over a climate apocalypse in less than a decade. Their answer is renewable electricity, and as this chapter has shown over and over, renewables cannot deliver reliable electricity to China, India, Africa, and other at-risk nations.[295]

Under current technology – and technological prognostications looking decades ahead – renewable electricity has no chance of ever reaching more than twenty-five percent of global electricity needs. Breakthroughs can happen, and both authors of this book, sincerely hope the sustainability aspect that is the wind and sun can be harnessed to make them deliver electricity effectively.

More importantly, for the poor, and this drive to alleviate a climate crisis by ridding the world of oil, petroleum, natural gas, coal, and nuclear is that a climate crisis doesn't exist. The climate and weather are always changing and have been doing so for four and a half billion years, but the facts show that man is not causing anthropogenic global warming due to fossil fuels, or nuclear energy.

This is deeply important, because if the west, and organizations with the money to fund electricity projects (western banks, investment firms, the UN, EU, U.S., World Bank, IMF, etc.) believe in climate change then poor peoples, nations, and continents will be left with unstable, intermittent, and costly solar and wind farms to meet their growing electricity needs. They will also have no access to all the products made from petroleum derivatives.

The facts say the earth's climate is "rising at a microscopically slow pace."[296] NASA's global temperature readings only go back to 1880, and since that time frame the earth's temperature went up 1.14 degrees Celsius. That averages out to an increase of 0.008 Celsius per year.

Miniscule when you compare this to prior geological periods when it was hotter, or cooler, and carbon dioxide was much higher.[297] Then climate models are clearly being shown to "project too much warming," and climate modelers "have a vested self-interest in convincing people that climate modeling is accurate and worthy of continued funding."[298] If taxpayer monies dried up would climate modelers even care about global warming/climate change?

Neither humans, nor fossil fuels were present during the five warming cycles that melted the icebergs of the five previous ice ages. Also, it should be noted, only twelve percent of the earth's surface is habitable by humans.

In recent years it was found to be some of the coolest on record.[299] And if it is warming, a warmer earth is found to save lives. The Lancet, a weekly, peer-reviewed, medical journal has "reported that worldwide, cold kills over seventeen times more people than heat."[300]

While the earth's temperature has risen 0.008 Celsius per year since 1880 then according to EMDAT, The International Disaster Database, "since the 1920s, the number of people killed annually by natural disasters has declined over eighty percent," this figure occurred, as the world's population went from two billion to over seven and a half billion.[301]

Emissions, and carbon dioxide increases have become the catch-all for anything wrong with the environment and global health, however, "global air pollution death rate fell by 50 percent since 1990 per 100,000 people."[302] The U.S. National Bureau of Economic Research (NBER) has negated the climate change will kill economies argument by estimating "if the earth's temperature rises by 0.01 degrees Celsius through 2100 – total U.S. GDP will be 1.88 percent lower in 2100."[303]

Meaning, if NBER calculations, and estimates pan out, then GDP per person is 178 percent higher. The U.S. Congressional

Budget Office has also validated these numbers, and by 2100 U.S. incomes will be almost triple per today's yearly paychecks.[304]

With those gains then technology, innovation, productivity, and longevity can mitigate barely hotter temperatures. Why would anyone advocate cutting electricity to China, India, Africa, and other developing nations over minimal climate change when the U.S. has shown that economic growth isn't affected if the temperature rises more than it has previously since 1880.

The biggest mistake over eradicating fossil fuels, and nuclear for developed, developing, and the world's poor comes from Climatologist Patrick J. Michaels who calculated:

> "That if the United States eliminated all carbon emissions – which would require Americans to give up fossil fuels, but also to stop breathing (to cease exhaling carbon dioxide) it would only reduce global warming by a negligible 0.052 Celsius by 2050.[305]

U.S. conservative think tank, The Heritage Foundation in U.S. Congressional research and testimony asserted the U.S. could shut down all energy and electricity usage, and global temperatures would still rise, because of China, India, and Africa.[306]

What we are finding is a nefarious movement within the U.S. government to subvert the electorates will towards energy, and elected President's.[307] Let's stop making the same mistakes as Al Gore, Tom Steyer, and Michael Bloomberg have along with all the other naysayers that global warming is the demise of mankind.

Let's focus on human progress, technological advancements, and pulling up our fellow humans stuck in beleaguered existences by giving them reliable electricity, and the six thousand products from petroleum derivatives. There is no danger of the world ending anytime soon over global warming, but we are in danger if we believe the climate-doom false prophets and extol the virtues of renewables.

Meanwhile, the poor among us burn wood, and cow dung to cook their food, and warm their families. We can do better, since we have the electricity technology that fossil fuels, and nuclear offer to save humanity.

CHAPTER FIVE

DEVELOPED COUNTRIES WITHOUT FOSSIL FUELS
Social changes to live in the pre-1900's

By Ronald Stein

INTRODUCTION

Many developing countries who are still stuck in the pre-nineteenth century's era have yet to join the industrial revolution. The prosperous societies in the developed, First World countries may need to implement many social changes to reduce the demands for the various derivatives from petroleum that are the basis of more than six thousand products in our daily lifestyle, and the fuels that support the demands of the various transportation infrastructures.

- The Green New Deal (GND) would NOT replace ALL energy generated by fossil fuels, JUST electricity by coal, natural gas, and nuclear.
- Even electricity cannot exist without fossil fuels as ALL the parts of solar and wind are made with the derivatives from petroleum.
- Wind and solar are incapable of providing all those derivatives from petroleum that are the basis of thousands of

products needed in every infrastructure, to "make things and move products"

- Electricity alone cannot support the military, airlines, cruise ships, nor merchant shops.

In case you don't remember, we also had virtually no military aircraft carriers, destroyers, submarines, planes and tanks around the world before 1900. Both WW I and II were won by the Allies, as they had more oil, petroleum, and coal than the Axis Powers of Germany, Italy, and Japan to operate their military equipment, move troop convoys, and supplies around the world.

The worlds' climate alarmism movement is trying to change current demands for fossil fuels, over to renewable electricity from wind and solar. Reducing our reliance on those deep earth mineral and fuels to zilch would require major social changes to live in the pre-1900's before oil began to support automobiles and airplanes; while making available the thousands of products from petroleum[308] that have enhanced the lifestyles of those in developed countries.

Our U.S. politicians' litigation and regulation grandiloquence seem to be unaware that oil and gas is not just an American business with its 135 refineries in the U.S[309], but an international industry with more than 800 refineries worldwide[310] that supply oil products and fuels to the world, The constant bickering to attract votes from those even less informed than our elected politicians seems to support their desires to hamper or close the American oil and gas industry.

Shockingly, the U.S. could literally turn off the entire country from any source of energy and global emissions would still grow according to U.S. Congressional testimony in 2017[311]. The entire U.S. economy, military and government could disappear, and global pollution, and respiratory illness would still rise. The reason why is "one of the biggest sources of carbon dioxide emissions is developing countries."[312] Think China, India, and Africa.

What we are seeing is a fossil fuel dilemma, where climate change, anthropogenic (man-made) global warming, and energy policies based on these premises are out to rid the world of carbon and greenhouse-gases since they are "a first-world problem."[313] The U.S., European capitals, and the United Nations (UN) seemed more concerned about the environment than the two billion people globally without the thousands of products from petroleum derivatives and electricity[314].

We're constantly hearing and reading about litigation against U.S. oil firms[315] like Exxon that they knew their industry was a potential harm to the environment. Many of the politicians are targeting lawsuits and further regulations against the oil and gas markets, and a ban on fracking to reduce the crude supplies to those U.S. refineries to hamper their production.

This endgame of ridding the world of thousands of products, affordable electricity, economic growth for Third World countries – even entire continents – is on the line without any answers[316] for how you replace fossil fuels and the thousands of products from petroleum derivatives with solar panels, wind turbines, and electric vehicles that just happen to depend on those same products from petroleum derivatives.

Access to affordable, plentiful, and reliable energy is closely associated with key measures of global human development including per-capita GDP, consumption expenditure, urbanization rate, life expectancy at birth, and the adult literacy rate. This research reveals a positive relationship between low energy prices and human prosperity. A similar level of human prosperity is not possible by relying on alternative sources of intermittent electricity from solar and wind power.

Interestingly, the 184 countries[317] that have ratified the Paris Agreement excludes Russia, Turkey and Iran[318] as those countries believe the countries that control natural gas and oil will control the world. Putin of Russia understands WW I and II were both won with fossil fuels to move planes, ships, tanks, troops and supplies.

SOCIAL CHANGES TO LIVE IN THE PRE-1900'S

Ideology says renewables will work, damn the consequences, the facts say otherwise. Abundant, affordable, reliable, scalable, flexible oil, coal, natural gas and nuclear electricity are why the planet is alive. Rid the world of this, and jobs disappear, economies wither, and modern, prosperous, peace-loving, sustainable nations collapse into anarchy. Upend the growth of fossil fuels globally, and watch China, India, Russia, Africa, and other growing nations undo the liberal-led order. Do this and lifestyles and "living standards spiral downward" for the remainder of this century[319].

In case you don't remember, we also had virtually no military aircraft carriers, destroyers, submarines, planes and tanks around the world before 1900. Both WW I and II were won by the Allies as they had more oil than the Axis Powers of Germany, Italy, and Japan to operate their military equipment and move troops and supplies around the world.

Before 1900, the world had very little commerce and without transportation there is no commerce. The two prime movers that have done more for the cause of globalization than any other: the diesel engine and the jet turbine, both get their fuels from oil. Road and air travel now dominate most people's lives.

Also, before 1900 the world had no medications, electronics, cosmetics, plastics, fertilizers, and transportation infrastructures. Looking back just a few short centuries, we've come a long way since the pioneer days.

Today, we've got 39,000 planes in the world[320] including all commercial and military planes. Yes, we've come a long way from the Wright brothers' historic first flight in North Carolina over 100 years ago. Yes, the same airlines that are accommodating 4 billion passengers per year[321].

We've got more than 60,000 merchant ships in commercial maritime transport[322] accounting for roughly ninety percent of world trade moving products around the world.

Regarding automobiles, earlier accounts often gave credit to Karl Benz, from Germany, for creating the first true automobile in the late 1800's, before Henry Ford sponsored the development of the assembly line technique of mass production. Today, there are 1.2 billion vehicles on the world's roads with projections of two billion by 2035[323].

Registration of electric vehicles is increasing, but only projected to be in the single digits, around five to seven percent. If projections come to reality by 2035, five to seven percent of the two billion vehicles would equate to 125 million EV's on the world's roads. The bad news is that would also represent more than 125 BILLION pounds of lithium-ion batteries that will need to be disposed of in the decades ahead.

The Energy Information Administration[324] (EIA) projects nearly a 50 percent increase in world energy usage by 2050, led by growth in Asia. Even with the rapid growth of electricity generation, renewables—including solar, wind, and hydroelectric power that is the fastest-growing energy source between 2018 and 2050.

The industrial sector, which includes refining, mining, manufacturing, agriculture, and construction, continues to account for the largest share of energy consumption of any end-use sector—more than half of end-use energy consumption throughout the projection period. The oil and gas industry are projected to remain the main player for world energy, so why are politicians so anxious to tank it?

The same U.S politicians that are thrashing on the oil and gas industry, and seeking its demise are the same ones reaping the benefits of the medications, medical equipment, communication systems, and electronics from the thousands of products from that industry that have contributed to their lifestyles and their ability to live beyond eighty years of age. They blatantly continue their guiltlessly use of planes, trains and vehicles to move them and their entourages around the world. Basically, they're enjoying everything they want to deprive from the poorest in developing countries.

The politicians' silence is deafening about deaths in poor countries that are mainly from preventable causes[325] of: diarrhea, malaria, neonatal infection, pneumonia, preterm delivery, or lack of oxygen at birth. Those underdeveloped locations in the world, mostly from oil and gas starved countries, are experiencing eleven million child deaths every year[326]. About 29,000 children under the age of five, or twenty-one each minute that die every day, And by the way, adults in those poor countries barely live past forty years of age.

Those children in poor countries still lack purified drinking water, sewage sanitation, adequate nutrition, reliable electricity (or any at all), adequate health care, i.e., the infrastructures and products we take for granted that are all based on deep earth minerals and fuels.

Seems that politicians supporting the demise of the oil and gas industry should speak up and take accountability for supporting the elimination of the industry that could reverse the annual fatality atrocities occurring in those poor countries, that are a few hundred years behind us and yet to even join the industrial revolution.

Could it be that our outspoken politicians that are aggressively trying to demise the U.S. oil and gas industry are unbeknownst supporting depriving the poverty-stricken in third world countries from the thousands of products we get from fossil fuels? The same list of products that have virtually conquered all the diseases and allows us to live in any weather condition. These products are now the basis of the lifestyles enjoyed only by those in developed countries, while the poorest in the world continue to suffer.

Without refineries, the U.S. economy would lose millions of jobs and take the financial hit to have all those products and fuels imported from foreign countries. The world will incur more emissions as most refineries outside our country, have significantly less environmental controls than here in our homeland. Our country would become a national security risk being dependent on foreign countries for our existence.

With or without the U.S. oil and gas industry that's the target of U.S. politician's discourse, the world will continue experiencing enormous fatalities of children worldwide in energy starved countries.

Do politicians want the oil and gas industry outside America supporting all the transportation needed for international commerce? Time to stop bickering and focus on what's important – saving millions of children from preventable deaths.

There are billions in undeveloped countries that have nothing to lose if the world ends, as they are living in the horse and buggy days that developed countries experienced centuries ago, since they have virtually "nothing to lose". They have yet to join the industrial revolution, and without oil and natural gas, they may never get that opportunity. They continue to live in poverty like they've been doing for centuries.

Almost half the world[327] -- over three billion people — live on less than $2.50 a day. At least eighty percent of humanity lives on less than ten dollars a day. More than eighty percent of the world's population lives in countries where income differentials are widening.

Incredibly, global poverty[328] affects the poorest forty percent of the world's population and they account for five percent of global income. The richest twenty percent accounts for three-quarters of world income. Water problems affect half of humanity.

So, what do we do with those developing countries that have yet to join the industrial revolution? The developed countries have enjoyed the benefits to human activities, lifestyles, and prosperity afforded by fossil fuels for the last couple of centuries.

Obviously, with limited income and poverty status, China and India with 2.7 billion people, continue to pursue the energy source that's abundant, reliable and affordable – coal. Those two countries are home to more than half (5,884) of the world's coal power plants[329] (10,210). Together they are in the process of building 634 new ones. They are putting their money and backs into their most abundant source of energy – coal.

Those developing countries yet to join the industrial revolution are missing out on several great things fossil fuels have done for humanity. Sunsetting the industry would negatively impact modern medicine, agriculture, longevity, and our ability to face natural weather disasters:

- Fossil fuels are the foundation of modern medicine and modern agriculture.
- Food abundance has increased even as the amount of land devoted to agriculture has declined, with former farm fields reclaimed by forests and pastures.
- Fossil fuels are saving lives and extending longevity now at eighty plus versus the forty plus range that we had just a few hundred years ago.

Countries where fossil fuels are in widespread use are generally wealthier and healthier, and prosperous countries are more resilient in the face of natural disasters than poorer countries lacking access to fossil fuels.

Developed countries have major ports for cruise liners that are accommodating twenty-five million passengers per year. Each liner consumes around 80,000 gallons of fuel PER DAY, per liner[330]._As a side note, billions of vehicle trips to and from airports, hotels, ports, and amusement parks are increasing each year.

For the billions of dollars of products imported and exported through ports, those 60,000 merchant ships that are moving products around the world are consuming more than 200 tons of fuel PER DAY, per ship[331] to move merchandise around the world.

There's something amusing about watching well-fed politicians and first world teenagers hit the streets to rail about fossil fuels, just before they're collected in stonking V8 Land Cruisers, whisking them away to their perfectly air-conditioned homes, jampacked with every electronic gadget, imaginable. The louder and more earnest their demands for an end to the use of fossil fuels, the sillier they sound.

In the U.S., Alexandria Ocasio-Cortez (AOC) is the Greenleft's favorite flibbertigibbet. Her New Green Deal – with its promise of 100 percent wind and solar delivering power for all, for free, and for all time – should have been dismissed as fairytale nonsense, long ago. However, there are plenty of pixie dream girls and boys happy to suck it up, like there's no tomorrow.

Ask them for details, and their responses range from evasive to delusional, disingenuous and outrage that you would dare ask. The truth is, they don't have a clue. They've never really thought about it. It's never occurred to them that these technologies require raw materials that must be dug out of the ground, which means mining, which they vigorously oppose (except by dictators in faraway countries).

Most have never been in a mine, oilfield or factory, probably not even on a farm. They think dinner comes from a grocery store, electricity from a wall socket, and they can just pass laws requiring that the new energy materialize as needed. And it will happen Presto!

It's like the way they handle climate change. Their models[332], reports, and headlines bear little or no resemblance to the real-world real world[333] outside our windows, on temperatures[334], hurricanes[335], tornadoes[336], sea levels, crops[337] or polar bears. But the crisis is real, the science is settled, and anyone who disagrees is a denier.

So, for the moment, let's not challenge their climate or fossil fuel ideologies. Let's just ask: How exactly are you going to make this happen? How will you ensure that your Plan A won't destroy our economy, jobs and living standards? And your Plan B won't devastate the only planet we've got? I'll say it again:

1. Abundant, reliable, affordable, mostly fossil fuel energy is the lifeblood of our modern, prosperous, functioning, safe, healthy, fully employed America. Upend that, and you upend people's lives, destroy their jobs, send their living standards on a downward spiral.
2. Wind and sunshine may be renewable, sustainable and eco-friendly. But the lands, habitats, wildlife, wind turbines, solar panels, batteries, transmission lines, raw materials, mines and laborers required or impacted to harness this intermittent, weather-dependent energy to benefit humanity absolutely are not sustainable and eco-friendly.
3. The supposed cure they say we must adopt is far worse than the climate disease they claim we have.

These climate alarmists seem to be oblivious to the fact that 100 percent of the industries that use deep earth minerals/fuels to "move things and make thousands of products" to support the economies around the world, are increasing their demand and usage each year of those energy sources from deep earth minerals/fuels, not decreasing it.

Electricity alone, especially intermittent electricity from renewables, is unable to support the energy demands of the military, airlines, cruise ships, supertankers, container shipping, and trucking infrastructures and thus has not, and will not, run the economies of the world, as electricity alone resulting from getting off fossil fuels would virtually negatively impact the following industries and infrastructures that are driven by the energy density of oil, coal, and natural gas:

- Medications and medical equipment that are all made with the chemicals and by-products of oil.
- Commercial aviation that has been accommodating four billion passengers annually.
- Cruise liners that have been accommodating more than twenty-five million passengers annually.
- Merchant ships that move products worth billions of dollars daily.
- The military presence that protects each country from each other, is increasing each year to save the world.
- Usage of fertilizers that accommodates growth of much of the food that feeds billions annually.
- Vehicle manufacturing as all parts are based on the chemical and by-products from fossil fuels.
- The usage of asphalt for road construction.

Getting-off-fossil fuels would reverse much of the progress society has made over the last few centuries. Until electricity storage technology can support intermittent electricity from wind and solar, the world will continue to have redundant fossil fuel backups for those windless and cloudy days to provide electricity to the world's economies around the clock.

The two prime movers of transportation, the jet turbine and the diesel engine, provide the backbone of our national prosperity, for both commerce and recreation. The demand for thousands of products refined from crude oil other than gasoline fuel—that is, the multitude of chemicals and by-products as well as aviation fuel, diesel fuel, and asphalt that the military and all infrastructures are dependent upon—directly supports our current lifestyles.

Modern, high-energy societies[338] enjoy a much higher standard of living than their historical predecessors, and these gains have naturally led to expectations of continued lifestyle improvements.

Products manufactured from fossil fuels have literally eliminated weather related deaths, are responsible for extended life expectancies, and have critically reduced fatalities of mothers and babies during childbirth. The search for a cure for cancer is partially driven by chemicals developed from and used in laboratories built by fossil-fuel by-products.

For the sake of argument, yes, getting-off-fossil fuels would reduce emissions, but such an abrupt stop that climate alarmists support would also drastically impact the lifestyles to which we've become accustomed and consequently move us backwards to medieval times.

The "green magic transition" may be the greatest con job in history.

Despite the hype over the ever-increasing connected capacity at wind and solar farms[339] worldwide, none, yes, let me repeat that, none have replaced any of the hydro, natural gas, coal, or nuclear generating plants that are providing continuous and uninterruptable electricity to people and businesses around the world.

Solar may work occasionally at homes and businesses as a source for supplemental intermittent electricity to lower daily demand from the grid, but they're still connected to a reliable source for continuously and uninterruptable power. We all know, if the sun is not shinning, their only source of electricity is the power generating plants feeding the grid even with the burgeoning mass storage technology popping up in the most auspicious places.

It's not that we're not trying to tap into the emission free electricity provided by Mother Nature, but wind and sunshine are too intermittent. They are not the panacea. They come with their own ills.

With the success the green movement has had on stymying nuclear, it's now attracting big oil companies to invest huge sums into the renewables craze.

There are three main reasons for that kind of investment from "big oil" into renewables[340].

- First, it's a great public relation move.
- Second, it's a fantastic business investment, as every wind and solar site generating intermittent electricity needs natural gas backup generating plants[341] to provide continuous and uninterruptable electricity.
- Third, if they fail, the government incentives are "no take back" guarantees and the loss is a tax write off. So, they basically get to dabble for free.

Another supposedly electrical solution to lessen our need for fossil fuels are electric vehicles, yet even a growth of one hundred times the number of EV's to 400 million by 2040 would only represent five percent of the global demand as total vehicles worldwide are projected to be two billion by 2035[342]. By some estimates, the total number of vehicles worldwide could increase to 2.5 billion by 2050.

In addition to the dismal impact projected for EV's is the dark side of their "green" technology[343], i.e., no supply chain transparency for the lithium-Ion batteries that power them.

The key minerals used in today's batteries are cobalt, of which sixty percent is sourced from one country, the Democratic Republic of the Congo (DRC), and lithium, of which more than fifty percent is sourced from the Lithium Triangle in South America, which covers parts of Argentina, Bolivia and Chile.

Today 20 percent of cobalt is mined by hand. The mere extraction of the exotic minerals cobalt and lithium used in the batteries of EVs presents social challenges, human rights abuse challenges, and environmental challenges. Amnesty International has documented children and adults mining cobalt in narrow man-made tunnels, at risk of fatal accidents and serious lung disease[344].

The cobalt mined by children and adults in these horrendous conditions in the DRC in Africa enters the supply chains of some of the world's biggest brands. There are no known "clean" supply chains for lithium and cobalt, yet the richest and most powerful companies in the world continue to offer up the most complex and implausible excuses for not investigating their own supply chains.

Elon Musk's electric product manufacturing conglomerate is one of the biggest offenders to provide transparency of the child labor atrocities and mining irregularities in the EV battery supply chain. Mr. Musk should read the laws, starting with The California Transparency in Supply Chains Act SB657[345] and followed by the U.S. with H.R.4842 - Business Supply Chain Transparency on Trafficking and Slavery Act of 2014[346].

Tesla Inc.'s "dirty little secret"[347] is turning into a major problem for the EV industry—and perhaps mankind. If you think Tesla's Model S is the green car of the future, think again. Energy independence, a reduction in greenhouse gas emissions and especially lower fuel costs promises are all factors behind the rise in the popularity of electric vehicles. Unfortunately, under scrutiny, all these promises prove to be more fiction than fact.

Another problem may be brewing for Tesla as some of their solar panels have been accused of initiating fires at Walmart[348] and Amazon[349] facilities. With lawsuits coming from two of the richest companies in the world onto the cash-bleeding Tesla, the gild is coming off the lily.

To drive the point home; while we in the developed countries with thriving economies continue to seek out an "alternative energy" that can maintain our lifestyles, the billions of people in undeveloped countries are starting to enhance their lifestyles with the most abundant and cost-effective energy source available to them today for electricity generation; coal.

When those billions rise out of poverty and develop modern economies, maybe, by then we'll have a better grasp on a real alternative to those deep earth minerals/fuels that renewable intermittent electricity cannot facilitate.

Granted we need to continue to pursue greater efficiencies and conservation in our daily lives. But, at the same time, we need to admit that the current state of green technology is just not working, as renewables have yet to replace continuous and uninterruptable electricity that people get from hydro, natural gas, coal, and nuclear generating plants.

CASE STUDY: CALIFORNIA

- *Germany was the first to go green and now has the highest cost for electricity, Australia was second and has the second highest cost. Now, California the 5ᵗʰ largest economy in the world, which has among the highest cost of fuels in America, is vying for the number three spot in the world for the cost of electricity.*

- *Even with the financial and environmental disastrous results in Germany over the last few decades, German Chancellor Angela Merkel vowed in 2019[350] that her nation would do "everything humanly possible" to curb the impacts of climate change. She isn't giving up and is willing to sacrifice the Germany economy to pursue her personal believes.*

- *With California's energy costs for electricity and fuels among the highest in the country, Governor Newsom just doubled down to increase inflation with actions to further reduce oil production and putting more electrical loads on a state that cannot generate enough electricity to meet its own needs. The California Governor's recent actions*[351] *will further "fuel" (no pun intended) the growth of the homeless and those on poverty.*

- *With its green dreams of an emission free state, California has not even been able to generate enough of its own electricity in-state and imported twenty-nine percent of its needs in 2018. The good news is that other state had the extra power. The bad news is that imported electricity comes at higher costs and those costs are being borne by residents and businesses alike. California households are already paying fifty percent more, and industrial users are paying more than double the national average for electricity.*

- *The future of electricity in California does not bode well either as the State has chosen to not challenge the closure of the States' last nuclear zero emission generating plant at Diablo Canyon and will be shuttering three natural gas generating plants in Southern California.*

- *With NO plans for industrial wind or solar renewable intermittent electricity projects to generate "replacement" electricity in-state for the shuttered plants there will be a need to import greater percentages from other states (if they can generate enough) to meet California's electricity needs in the years ahead. As you may know, the public has been underwhelmed with the huge land requirements for those renewables, so future large wind and solar sites are becoming less likely. And as you guessed it, more costs to the consumers and businesses who are already infuriated with high costs.*

- *Governor Newsom should know California is the only state in the continental United States that currently imports most of its crude oil energy from foreign countries. The California Energy Commission*[352] *(CEC) data demonstrates that this dependency on foreign sources of oil requires expenditures of sixty million dollars EVERY*

DAY to oil rich foreign countries to support the fifth largest economy in the world for it's military, aviation, cruise ships, and merchant ships, just to make up for the States' choice to continue decreasing in-state production.

- *California has chosen to be the only state in America that imports most of its oil needs from foreign countries and relies on the U.S. Navy[353] to pay a steep price keeping an aircraft carrier with escorts on station to deter attacks on oil tanker traffic operating in and around the Persian Gulf.*

- *The Governor seems to be oblivious to the fact that one hundred percent of the industries that use deep earth minerals/fuels to "move things and make thousands of products" to support the economies around the world, are increasing their demand and usage each year of those energy sources from deep earth minerals/fuels, not decreasing it.*

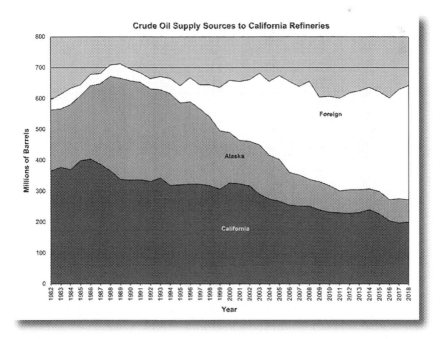

- *California's love of foreign crude oil is obvious as California increased crude oil imports from foreign countries from five percent in 1992 to fifty seven percent in 2018[354].*
- *The CALIFORNIA DELUSION[355] YouTube video by Mark Mathis of the Clear Energy Alliance that's gone viral with more than fifty-six thousands views discusses California's love for imported crude oil to meet the state's energy needs for military, airlines, cruise ships and merchant ships, trucking and automobiles that's putting America at a national security risk.*
- *California's ban on oil drilling and fracking will guarantee the growth of the homelessness, poverty, and welfare populations, by fueling (no pun intended) the housing affordability crisis. Life as we know it will change as our advanced lifestyle reverts to medieval times.*
- *California has been producing oil for over 100 years from more than 80,000 active and inactive wells.*
- *Sacramento Democrats are supporting Assembly Bill AB-345 (Muratsuchi), "Oil and gas: operations: location restrictions," which would require, commencing January 1, 2020, all new oil and gas development outside federal land, to be located at least 2,500 feet (nearly half a mile) from any residence, school, childcare facility, playground, hospital, or health clinic. The bill would define re-drilling of a previously plugged and abandoned well, or other rework operations, as a new development.*
- *Today, there are almost 10,000 active or newly permitted oil and gas wells[356] located within a 2,500' buffer of sensitive sites, that represents 13.1 percent of the total 74,775 active wells in California. There are another 6,558 idle wells of nearly 30,000 idle wells but putting these idle wells back online would be blocked if the wells require reworks to restart or ramp up production.*

- *The Governors latest moves to reduce production and require larger setbacks for existing production wells will further decrease production and require the State to increase its monthly imports resulting in expenditures approaching a whopping eighty million dollars EVERY DAY, from the States' current cost of sixty million dollars EVERY DAY for foreign countries to support our infrastructures.*

- *AB 345 will simply result in increased crude oil imports from foreign sources not operating under the same precise environmental standards as California and will lead to significantly higher transportation costs and an increase in greenhouse gasses and other emissions associated with bringing that oil into the state, as no other state or country has the stringent environmental controls as California.*

- *Does the Governor and Assemblyman Muratsuchi know AB 345 actions are supportive of California continuing to increase the cost of fuels to its forty million residents, and further solidifying the fifth largest economy in the world being a National Security risk to America?*

- *Its mind boggling that our California legislative leaders continuously fail to see the direct correlation between high energy costs for electricity and fuels, and poverty, homelessness, and a housing affordability crisis already impacting the Golden State.*

- *There are scary similarities between Governor Newsom's goals for California and Vladimir Putin's objectives. Both support California being more and more dependent on imported foreign oil, and both support anti-fracking in California as a successful fracking enterprise would lessen the states' dependency on that foreign oil. Does the Governor know his actions are supportive of California becoming a National Security risk to America?*

- *The charge into green will require retraining of a huge displaced workforce used to a certain lifestyle. A minimum wage earner is not afforded the time nor the resources to enjoy leisure activities. Their mainstay in life is to make ends meet. Most times that requires two and three such jobs and maybe even both heads of household working two of those low wage jobs to break even financially. How does the Governor plan to feed the families of displaced workers when he shuts down their means of survival? By default, his actions will increase the welfare numbers.*

- *Regarding the Governor's move to require EV's to replace state fuel driven vehicles, I have no problem having the state buy EV's but I have a major problem with the Governor "blowing off" the transparency of the child labor atrocities and mining irregularities in the EV battery supply chain.*

- *The Governor should read the laws, starting with The California Transparency in Supply Chains Act SB657[357] and followed by the U.S. with H.R.4842 - Business Supply Chain Transparency on Trafficking and Slavery Act of 2014[358]. The richest most powerful companies in the world, and now the Governor of California are still making excuses for not investigating the supply chains and continue to power manufactured EV's with "dirty batteries".*

- *Like the green movement effects on the economies in Germany and Australia, the governor's plan of moving forth at such an abrupt pace to end the states' dependence on fossil fuels and convert to one hundred percent renewable electricity (folks, it's not renewable energy, it's only intermittent electricity) will break the back of the oil industry in the state and severely damage the California economy.*

- *Like Germany[359] and Australia[360], and their experiences with disasters to their economy and environment from the green movement, California is following, not leading as they would like you to believe, into known disastrous territory. Just like Germany and Australia before trudging into the green morass, our leaders cannot "see' the direct correlation between energy costs for electricity and fuels, and homelessness and poverty.*

- *Efficient energy systems affect everything, not only from transportation, but the cost of groceries and food and cleaning products. For the working class, after fuel and electricity costs, what's left in the purse, if there is anything left, goes toward the other living expenses. Lately, there's been less and less left.*

- *While the state's crusade toward 100 percent renewables continues, the intermittent electricity resulting from all those wind and solar farms ("100 percent by 2045")[361] has caused the state to increase its imports of crude oil to meet its insatiable energy demands. California could decrease its growing dependency on foreign crude oil simply by increasing in-state crude oil exploration, but environmentally correct state leadership refuses to even to consider such a plan.*

- *High cost of electricity[362]: California's electricity is already fifty percent higher than the national average for residents, and double the national averages for commercial, and are projected to go even higher.*

- *The inability to replace the closure of continuously uninterruptable electricity from nuclear and natural gas with renewables of Industrial wind and industrial solar is causing the state to import more and more of its electricity. Without the huge land requirements for industrial wind and industrial solar renewable electricity, the need to import more will escalate every year.*

- *And now, starting July 1ˢᵗ another six cents is being added to the posted price at the pump for infrastructure repair and maintenance. With residents already paying as much as an extra dollar for fuel, we should already have the best roads.*

- *More insults are coming from Washington DC. Senator Debbie Stabenow (D-Mich) is sponsoring legislation to raise the ceiling for tax credits to EV from 200,000 to 600,000 electric vehicles per manufacturer and continue the $7,500 tax credit to buyers that can afford to buy their own cars without it. This extension of the tax credits to the wealthy is grossly unfair to middle- and lower-class consumers who are unable to afford electric cars.*

- *The main supporters of extending the lucrative tax credits to more EV's are the dozens of automakers who want to sell cars, several major environmental and public health groups, the Sierra Club, and electric utilities that want to sell electricity. The middle- and lower-class consumers are not represented in this proposed legislation at all.*

- *Again, it's the ninety five percent of the residents who can't afford electric vehicles who will pay for those tax credits and road taxes the EV owners are not paying for because they don't go to the gas pumps where the taxes are levied.*

- *California is doing everything right with the green crusade to increase costs of electricity and fuels which guarantees growth of homeless, poverty, and welfare. It's scary that our leaders can't "see" that the regressive energy policies have serious consequences for working families. Their misguided directives are intertwined with every aspect of daily life and is causing the continuous growth of poverty and homelessness from the Oregon state line on the north all the way to the Mexican border on the south.*

If we continue to ban mining under modern laws and regulations here in America, those materials will continue to be extracted in places like Inner Mongolia[363] and the Democratic Republic of Congo[364], largely under Chinese control – under labor, wage, health, safety, environmental and reclamation standards that no Western nation tolerates today. There'll be serious pollution, toxics, habitat losses and dead wildlife.

Even worse, just to mine cobalt for *today's* cell phone, computer, Tesla and other battery requirements, *over 40,000 Congolese children* and their parents work at slave wages, risk cave-ins, and get covered constantly in toxic and radioactive[365] mud, dust, water and air. Many die. The mine sites in Congo and Mongolia have become vast toxic wastelands. The ore processing facilities[366] are just as horrific.

Meeting GND demands would multiply these horrors many times over. Will Green New Dealers require that all these metals and minerals be responsibly and sustainably sourced, at fair wages, with no child labor – as they do for T-shirts and coffee? Will they now permit exploration and mining in the USA?

Meeting basic ecological and human rights standards would send GND energy prices soaring. It would multiply cell phone, laptop, Tesla and GND costs five times over. But how long can Green New Dealers remain clueless and indifferent about these abuses?

Up to now, this has all been out of sight, out of mind, in someone else's backyard, in some squalid far-off country, with other people and their kids doing the dirty, dangerous work of providing essential raw materials. That lets AOC, Senators Sanders and Warren, Al Gore, Tom Steyer, Michael Mann, Greenpeace and other "climate crisis-renewable energy" profiteers preen about climate justice, sustainability and saving Planet Earth.

Understanding energy will help us formulate the best energy policies to bring billions of people out of crippling poverty, lowered life expectancies, and even protect them from the allure of terrorism. There is an opportunity to alleviate poverty and war if every person on the planet can be given access to scalable, reliable, affordable, abundant, and flexible energy. When basic facts are ignored or shoved aside for political gain, however, rational energy discussions no longer happen.

I presume the alarmists that constantly refuse to surface from behind their tweet machines to debate is because they have no case to debate the facts that they are using to justify their growing alarmist vocabulary. Unless there's a face to face debate with the supposedly deniers, that have more data than words, we'll never hear both sides of the climate discussions. Looking forward to face-to-face discussions.

CONCLUSION

Before 1900, life was hard and dirty, with many weather and disease related deaths, and life longevity that was only in the forty plus area. Once petroleum proved to be the energy source that could meet the demands of society for fuels, and the thousands of products made from the derivatives from petroleum, the world changed dramatically and is in constant state of technological changes.

Here's a partial list of the infrastructures and products that would become scarce or non-existent with the GND, resulting is catastrophic social changes that would be necessitated in developed countries to adjust to the lifestyles that existed before the inventions of the automobile and the airplane.

- NO military equipment: aircraft carriers, battleships, destroyers, submarines, planes, tanks and armor, trucks, troop carriers, weaponry
- NO medications and medical equipment
- NO vaccines
- NO fertilizers to help feed billions.

- NO cell phones, computers, and I Pads
- NO vehicles
- NO airlines that now move four billion people around the world
- NO cruise ships that now move twenty-five million passengers around the world
- NO merchant ships now moving billions of dollars of products monthly throughout the world
- NO tires for vehicles
- NO asphalt for roads
- NO water filtration systems
- NO sanitation systems
- NO space programs

The worlds already experienced life without fossil fuels as recently as a few short centuries ago. We never had the oil industry before the 1900's, so imagine how life will be like with no infrastructures to move things that are the basis of commerce, and no chemicals to make the products that are the basis of our lifestyles?

CHAPTER SIX

BILLIONS WITHOUT BASIC ELECTRICITY
Affordability will be the future

By Todd Royal

INTRODUCTION

To meet the demands of society, the United States Energy Information Administration (EIA) expects U.S. oil production to average approximately 13.3 million barrel per day (mb/d) in 2020, and continue growing into the foreseeable future.[367] This is a nine percent increase from 2019, and has been steadily growing since the U.S. shale revolution began in 2006-7. This evolution is being led by relative newcomers to the shale revolution – the U.S. state of New Mexico – which "hit one mb/d of oil production in November, according to international energy consulting firm, Rystad Energy."[368]

Carbon-free electricity took a positive turn when the U.S. Congress in late December 2019, "voted to approve appropriations for fiscal year 2020 that includes $1.5 billion for nuclear energy programs, a twelve and a half percent increase from 2018." Maria Korsnick, President and CEO of the Nuclear Energy Institute (NEI) said, "nuclear energy is an essential driver in lowering carbon emissions (and) a cleaner environment."[369]

Nuclear generated electricity is the only zero-carbon source of electricity.[370] It lowers overall carbon emissions while not producing any air pollution or respiratory illnesses.

Great news abounds for future customers, nations, and continents over these developments to bring affordable, reliable, and basic electricity to end-users around the world. When energy grows – in all forms – particularly, fossil fuels, and nuclear generated electricity then human prosperity, longevity, and innovation increases exponentially.

But electricity is under attack, and this has severe, deadly consequences for the two billion people globally who do not have access to reliable electricity.[371] Over 600 million Africans don't even have basic electricity; they are literally, living in the Dark Ages.[372] Leading, U.S. Democratic Socialist candidate, Vermont Senator Bernie Sanders has promised to ban almost anything that will allow electricity to be produced. He wants to nationalize the electrical grid and claims electricity would be virtually free by 2035.[373]

The figure below illustrates how the U.S. is growing in monthly crude oil production, and how the rush into alternative energy forms without taking into consideration this upward fossil fuel growth pattern. This higher movement leads to surging job growth in all sectors of the globally connected economy, and greater tax revenues for local, county, state, and federal governments.

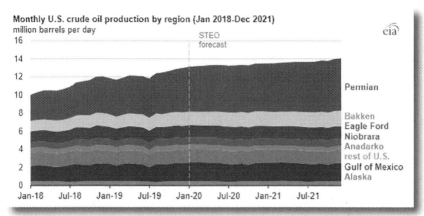

Monthly U.S. crude oil production by region (Jan 2018-Dec 2021)
million barrels per day

STEO forecast

Permian

Bakken
Eagle Ford
Niobrara
Anadarko
rest of U.S.
Gulf of Mexico
Alaska

Source: U.S. Energy Information Administration, *Short-Term Energy Outlook*, January 2020

Further troubles come from the world's largest money manager, BlackRock, announcing they are "planning a (energy to electricity) storage fund, (and) is creating a multibillion-dollar renewable energy fund, which follows on the heels of its decision to no longer finance coal projects."[374] BlackRock believes they are being noble, and they will lower emissions by taking away the most reliable source of electricity – coal-fired power plants.

Currently, 1,600 coal-fired power plants are either being planned, or under construction in sixty-two countries.[375] If BlackRock, or Sanders believes they are saving the planet they don't seem to understand that people, towns, villages, cities, counties, states, nations, and continents want reliable electricity. Saving the planet doesn't cross their minds.

BlackRock doesn't realize, or doesn't care that unless taxpayer subsidies underwrite entire wind, solar, and electricity storage projects they will fail – like the German government (more on Germany later in this chapter) has by attempting to transition to renewables, and energy storage systems.[376] Thus, securing fewer monies for reliable continuous, uninterruptable electricity that is generated from coal, natural gas, and nuclear generation.

Hundreds of billions of development money is being wasted on renewables, instead of fossil fuels, or nuclear generated electricity for billions without hope, or a future.[377] Even countries like Great Britain are saying they will ban all fossil fuel/internal combustion vehicles before giving any specific plans for how that will take place?[378]

Much less if their electrical grid can handle intermittent, spiked electricity that comes from wind turbines and solar panels; or if the grid can handle tens of millions of electric vehicles charging at the same time? Under current technological, and future scenarios that type of grid has not even come close to being invented yet. Britain will also need more electricity to make their entire transportation sector electrical. A new electrical grid will need to be built.

The math is the most interesting part of the British, American, and any other country considering going all-electric for their nation's transportation sector.

Just matching the electrical requirements for a typical gasoline-filling station, an all-electric station "would have to have thirty megawatts of capacity, equivalent to the electricity use of 20,000 homes."[379] This new all-electric station would then need "600 of those 50kW (kilowatt) chargers for a station to service 2,000 cars in a twelve hour time frame."[380]

This conservative estimate "would require a (approximately) twenty-four-million-dollar investment just for the cheapest rechargers." If the station is busier it could need enough electricity to power over 75,000 homes – "and that's another (approximately) thirty million dollars. If you want to go the windmill route, you'll need ten of them, each costing roughly four million dollars." Those ten windmills/turbines are only for one charger/former gasoline pump.[381]

Wind turbines are so grossly inefficient for the British that for over ten years:

"British wind farms have received constraint payments to reduce their output because of electricity
grid congestion. A total of 649 million British
Sterling Pounds have been paid out since 2010
for discarding 8.7 TWh of electricity. To put this
into context, this quantity of energy would be
sufficient to provide ninety percent of all Scottish
households with electricity for a year."[382]

All this electrical usage and charging or discarding of unused
renewable electricity would likely take place during peak-usage,
daylight hours, and unless mandated by law to only charge at
night.

This British policy of attempting to go all-electrical is similar
to a "majority of U.S. adults believing climate change is most
important issue today," without having any idea what the world
looks like without fossil fuels?[383] Both countries are moving towards an unsustainable, and national security risk by eliminating
fossil fuels, and hurting people all over the world who needs
electricity and global security the west provides.

Russian energy conglomerate, Gazprom, on the heels of the
BlackRock announcement said, "they will move forward without
foreign companies in order to complete The Nord Stream 2 pipeline," defying U.S. sanctions against the Kremlin.[384] The main
client(s) for Russian natural gas will be Germany, and Western
Europe whose main mode of protection against Russian influence, or invasion is U.S.-led NATO.

Europe is aiding Russia taking billions from European citizens, and militaries when Russia uses those funds to continue
the weaponization of energy that is oppressing millions in Syria,
Russia, Ukraine, Central Asia, and the Middle East.[385]

This geopolitical and western push-and-pull against fossil fuels and nuclear (the Bernie Sanders camp) versus China,
India, Africa, Russia, and other countries not under the western,

security-alliance umbrella (NATO, ASEAN) is real, and growing daily. Caught in the middle are two billion people without
electricity.

Add bank executives, institutional investors, and political
figures interested in heavily subsidized wind and solar farms
for intermittent electricity, and a conundrum unfolds. One that
excludes fossil fuels, and nuclear to the detriment of our planet.
This will cost trillions in unproven technologies that wind and
solar provide.[386]

Either continue investing in, and building coal-fired, natural
gas-fired power plants, and nuclear generating electricity plants
that fulfills base-load generation capacity, or build and invest in
unreliable, intermittent, expensive, and mathematically unstable
wind and solar farms.[387]

Everywhere, from Texas to Germany and Australia, that
renewables have been deployed electricity prices can spike over
40,000 percent.[388] That's correct, a 40,000 percent price spike.
Base-load requirements are only met at this time by fossil fuels,
and nuclear under current energy and electricity technological
constraints to meet the basic standards of energy and electricity
being abundant, affordable, scalable, reliable, and flexible.

We are worshipping the sustainability of the wind and the
sun without considering the billions who are in abject, agrarian
poverty since they do not have electricity. The 600 million in
Africa without reliable electricity is one of the biggest problems
the world has right now, and it will only become thornier years,
and decades ahead.

This problem is highlighted by religious persecution, and
the slaughter of Christians taking place at the hands of Islamic
militants in Nigeria.[389] If the entire African continent, and particularly Nigeria, had reliable electricity this problem doesn't have
room to fester, and mushroom into brutal slayings of religious
minorities.

The world needs electricity, and an opportunity to join the

modern world, advanced by western interests for over 150 years with the advent of the industrial revolution.

FOSSIL FUELS & NUCLEAR POWER THE WORLD

Full steam ahead – that is the unsaid mantra of China and Japan – who have left intermittent wind, and solar farms behind to "build hundreds of New-Age Coal-Fired Plants."[390] The unintended consequences of renewables are having the opposite effect of saving the planet: coal-fired power plants, which are reliable, energy-dense, scalable, affordable, and flexible are being built at such a rapid clip that any global, western-led environmental gains are being negated over the unreliability of renewables.

According to the International Energy Agency's (IEA) *World Energy Outlook 2019*, said:

> "While the importance of transitioning to a carbon-neutral world is recognized, society is still heavily dependent on fossil fuels. Economic growth and a rising global population means that renewable energy sources can't keep up with worldwide energy demand."[391]

Countries, nations, and continents understand the term, "self-interest rightly understood," and reliable electricity is a top national security concern the same way a nuclear arsenal is for first, second, and third world nations.[392] It's why the great power China is keeping up with Japan, and building hundreds of coal-fired power plants.[393] Wind and solar have fallen out of favor for both countries. Coal doesn't need perfect weather to deliver continuous and un-interruptible electricity 24/7/365.[394]

And why wouldn't both countries build coal-fired plants, and use them? When global environmentalist, Bjorn Lomberg, who wholeheartedly believes in global warming and climate change, says:

> There are "Global priorities bigger than climate change," then both Asian powerhouses, and the billions without electricity are going to use anything possible (coal, natural gas, oil, petroleum, nuclear, future technologies) to acquire reliable electricity.[395]

The entire Asian hemisphere is growing; does any sensible, rational, non-ideological person believe Asia is going to de-carbonize by de-industrializing? Fossil fuels will also rule in Africa, which is expected to add "more than a half a billion people (to its) urban population by 2040."[396]

To put this into context, China didn't add this many people during its population burst between 1990-2010. During this time frame "China's production of materials such as steel and cement skyrocketed (both materials rely heavily on reliable electricity and crude oil)."[397] Industrial electrification is what billions in Africa and Asia need – besides residential electricity. The solution for renewables to be the only source of electricity is to:

> "Remove (taxpayer) subsidies, mandates and other forms of energy favoritism by letting technologies advance or fail in markets based upon their own true merit or lack thereof."[398]

Otherwise, billions will have negative consequences from an environmental-cabal, hell-bent on de-carbonization, and the solution coming from the wind and the sun. Consumers everywhere will have the "consequences of various policy regimes denying consumers choice to best satisfy their energy usage requirement."[399]

The De-carbonization movement is killing electrical generation for non-advanced countries, and continents.[400] The costs are
over ten trillion dollars annually. Here is a basic explanation for
why countries such as China, India, Japan, and others use fossil
fuels. Besides the over six thousand products that come from the
derivatives of petroleum, fossil fuels are known energy-dense
quantities. Energy density is defined as:

> "The energy stored per pound (and) is the critical
> metric for vehicles and, especially, aircraft, cruise
> ships and merchant ships. The maximum poten-
> tial energy contained in oil molecules is about
> 1,500 percent greater, pound for pound, than the
> maximum in lithium chemistry. That's why (all)
> aircraft, cruise ships and merchant ships, and
> rockets are powered by hydrocarbons."[401]

Energy density is one of the biggest reasons, outside of cost,
why electric vehicles (EV's) won't work when coupled with eliminating fossil fuel-based motor fuels "in favor of renewable-based
electricity."[402] EV's under current technological constraints are
operating as, "emission elsewhere" vehicles.[403]

Fossil fuels, and nuclear generated electricity are known physical quantities and qualities. Their favorable thermodynamic
properties are the reasons why growing, and advanced countries
will likely choose oil, petroleum, natural gas, coal, and nuclear.

Human wishes and desire-fulfillment economically, and historically began when fossil fuels, and nuclear energy were incor-
porated into everyday life. Nuclear was after the Second World
War, and is desirable, because it contributes to zero-carbon,
electrical emissions. Both working together give certainty
whereas "renewables pose a real threat to conventional sources
of electricity."[404]

It is why since 1990 the global total primary energy supply

has mainly been derived from fossil fuels, according to the International Energy Agency (IEA).[405] Billions need persistent, reliable, consistent sources of electricity that fossil fuels provide – even more so – than nuclear generated electricity.

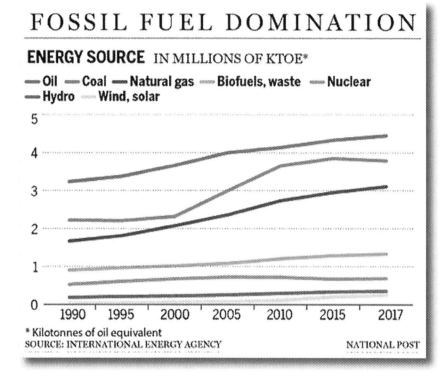

FOSSIL FUEL DOMINATION

ENERGY SOURCE IN MILLIONS OF KTOE*

— Oil — Coal — Natural gas — Biofuels, waste — Nuclear
— Hydro — Wind, solar

* Kilotonnes of oil equivalent
SOURCE: INTERNATIONAL ENERGY AGENCY   NATIONAL POST

If fossil fuels power and electrify the world, then American fracking has changed the world in every conceivable way related to electricity. The United States is expected sometime in 2020 to become a "net exporter of energy."[406] An astonishing development – when you consider – during the President George W. Bush administration the U.S. imported most of its oil and natural gas for domestic use.

China, India, Africa, most of Central America, the Middle East, and large parts of Central Asia are struggling with little to no electricity. How can anyone from the west, U.S.

environmentalists, and western environmental organizations, believe all the above-mentioned nations, and continents won't follow the U.S. model for fossil fuel use?

Why won't all of them – especially China, India, and Africa whose populations are growing the fastest – plan for, allow, and legislate their economies, and national security reach new heights by fracking, or other methods of exploration and production (E&P) for fossil fuels?

The 2019 United States Energy Information Administration's *International Energy Outlook* factually reveals China, India, and Africa are using more fossil fuels and nuclear generated electricity to power their energy, and electricity hungry billions of citizens more than ever before.[407]

THE CONFUSED WEST LED BY THE U.S.

The U.S. and its largest state, California, highlights how America is a divided nation when it comes to electricity. If you simply examine China, they are never going to follow the destructive, ineffective, California environmental model for electricity, which is atrocious for economic vitality.[408]

China understands what painful, sub-human, agrarian poverty looks like better than anyone in the last 150-200 years. More importantly, China along with India, and Africa quickly understands this reality – what life is like without fossil fuels, and the over six thousand products that originate from petroleum.[409] It is wretched, putrid, disease-filled, and leads to lower birth rates, and early deaths.

Other nations and continents leaving abject poverty without electricity realize California, and large parts of the U.S. buying into green new deals, renewable futures, and zero-carbon societies are left with: "The dystopic reality of mass homelessness, filth and rampant inequality that increasingly characterize these (U.S.) urban cores."[410]

The Green New Deal (GND), which none of these nations, or continents will ever implement was never about fighting global warming, saving the earth, or the environmental catch-all, known as climate change. Christiana Figueres, executive secretary of the United Nation's Framework Convention on Climate Change, laid out the true intent when she admitted:

> "The climate-fear campaign was an instrument for replacing capitalism (fossil fuels, reliable electricity, and nuclear generated electricity) with a more socialist, centrally planned economy. (And) This is the first time in the history of mankind that we are setting ourselves the task of intentionally, with a defined period of time, to change the economic development model that has been reigning for at least 150 years, since the Industrial Revolution."[411]

Naïve destruction at best, and global security at worse is at stake when it comes to energy and electricity for billions will only rely on intermittent wind, and solar electricity for their militaries, national security, and survival.

China, India, or even African nations such as Nigeria battling militant Islamist group Boko Haram will never follow western energy, electricity, and environmental policies.[412]

Environmental movement(s) in all forms: individual, corporate, non-profit, and government, whose original existence was to save dirty waterways, unhealthy air, and shorter life spans has taken on a new, insidious tone. Conservation coupled with ecological purity is thrown aside for raw power, and the political gain that environmental socialism offers.

This uses western, leftist political ideologies cloaked in climate change-nihilism whose main beneficiaries are the U.S. Democratic party, Western European green groups, and environmental foundations ironically enough like the Rockefeller Foundation, which made trillions from oil and petroleum.

Billions without electricity are left with nothing. None of this makes sense to an African, Chinese or Indian from India why they would implement these un-effective environmental policies from say – Massachusetts Senator, Elizabeth Warren who said, "By 2028, no new buildings, no new houses (would be built) without a zero-carbon footprint," if she ever become the U.S. President.[413]

Senator Warren has not stipulated how this would take place since every part for how buildings are constructed (steel, concrete, rebar, nails, aluminum siding, etc.) originates from a barrel of crude oil broken down in a petrochemical plant.

U.S. Socialist Senator Bernie Sanders believes "we need national rent control," to make housing more affordable in the U.S. and other parts of the world.[414] California has the highest median home prices in the U.S., and some of the highest in world, illustrating when housing is expensive, energy and electricity is expensive, and vice versa.[415] Housing and energy prices complement one another negatively and positively as well.

What offers the most damage against electricity being delivered globally is when the environmental movement stays silent on destruction in all forms from countries such as China and Iran. Historian and classicist, Dr. Victor Davis Hanson says:

> "If the West is guilty of carbon crimes, racism, and bigotry, what are China and Iran? Greta Thunberg (the Swedish teenager, and green global celebrity) might be more effective in her advocacy for reducing carbon emissions by redirecting her animus. Instead of hectoring Europeans and Americans, who have recently achieved the planet's most dramatic drops in the use of fossil fuels, Thunberg might instead turn her attention

to China and India to offer her 'how dare you'
complaints to get their leaders to curb carbon dioxide emissions."[416]

Whether or not China, India, and Iran loosen their grip on
totalitarianism, or lessening their CO2 emissions will depend on
their leaders: not on a confused, mentally ill teenager. More importantly, China and India "account for over a third of the global
population and continue to grow their coal-based industries."[417]
The Nobel Prize Committee should be ashamed for nominating
Greta Thunberg for a Nobel Peace Prize.[418]

This little girl (Greta Thunberg) needs guidance, love, and
parental acceptance (not directed manipulation) while furthering
her education. Please realize that Greta, and other misguided
teenage climate protestors have the mindset of a young person
whose prefrontal cortices, "which regulate decision-making,
planning, self-awareness and inhibition, do not fully develop
until we (human beings) are in our mid-twenties."[419]

To say manipulated teenagers are the case study in non-wisdom
thinking is an understatement. They will leap at the first chance
to cause chaos and have zero understanding of long-term planning or consequences of their actions. Reckless behavior is the
norm, and if scientific analysis, econometrics, regressions coupled
with climate modeling is their goal, then it hasn't been stated.

Emotional outbursts, and a chance to upset the Presidency of
Donald Trump seem to be their aim. Protests are not legislative
changes that have any lasting inspiration the way the international treaty – Montreal Protocol on Substances that Deplete
the Ozone Layer did by cleaning up and phasing out production
of CFCs (typically gases used in refrigerants and aerosol propel-
lants) "responsible for ozone depletion."[420]

This was an involved collaboration between elected officials, policymakers, scientists, industry leaders, and public
transparency. What this chapter, and book would ask young

Ms. Thunberg, but more importantly, the adults using her for their own gain is a climate model simulation brought forth by Astrophysicist Willie Soon, formerly of NASA.

Mr. Soon believes there are twelve variables such as "differential rotation between the earth's surface and earth's core, and the entire solar system's magnetic field and gravitation interaction," while admitting, "these twelve variables are not well understood, and there are thousands that affect future temperatures and climate."[421]

Additionally, just the U.S. government has poured billions into mathematical modeling, and climate predictions that have not "accurately predicted anything as to our climate over the past 30 years."[422] Most damning for Ms. Thunberg, child climate protestors, and the adults taking advantage of them for votes, and financial gain is this from Mr. Soon:

> "With these facts in mind it is no surprise that if we actually knew all the variables involved in a reasonable mathematical climate model it would take a supercomputer forty years to reach an answer to each question we posed."[423]

If Ms. Thunberg continues her misguided, woke-quest then she should start with Iran, Russia, China, and India whose coal-based industries, and coal-fired power plant electrical generation capacity is growing daily.[424]

Whereas, U.S. fossil fuel production is booming with fewer wells in production, would Ms. Thunberg want Iran, China, and India to produce fossil fuels instead?[425]

With over six thousand products that have their beginnings from a barrel of crude oil, this little child, and her minions wouldn't survive without crude oil. What is her answer to this demand dilemma from society for modern products that come from deep earth minerals?

U.S. crude oil and natural gas production increased in 2018, with 10% fewer wells

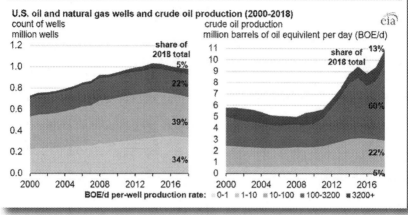

Let Greta, and other environmentalists advocate for safe, zero-carbon, reliable electrical power generation that comes from nuclear generating plants. The French receive roughly "eighty-five percent of their power from nuclear power plants."[426]

It is the only source of emission-free electricity in existence. But public perception over industrial mishaps like "The Three Mile Island accident, (which) caused no physical harm, but the event changed the public perception of the risks of nuclear energy."[427] This misguided perception hampers the ability to deliver electricity to billions.

From the outset, there has been a strong awareness of the potential hazard of both nuclear criticality and release of radioactive materials from generating electricity with nuclear power. As in other industries, the design and operation of nuclear power plants aim to minimize the likelihood of accidents and avoid major catastrophes when they occur.

Nuclear related deaths: Worldwide total (not annually, but from inception of nuclear) nuclear deaths including Three Mile Island (March 1979),

Chernobyl (April 1986) and Fukushima (March 2011) are LESS than 200.

To put the above numbers into perspective, of the millions that die each year from starvation, diseases, weather, air pollution, driving, working, walking, and overdosing, nuclear related deaths have been less than 200 worldwide, not annually, but from inception of the industry.

Where does Ms. Thunberg stands on these facts? Both authors of this book deeply feel for Greta Thunberg, and where she ends up in life; and only want the best for her. We hope she uses her unfounded celebrity on advocating for billions without electricity, and the hope for fossil fuels and nuclear to expand national, continent-wide, and global human prosperity.

What kills billions of lives from not reaching their full potential is when western policymakers, well-known environmental organization, and influential environmentalists embrace policies like Germany.

Germany's "Energiewende (transition to green energies) is driving up prices," and while wealthy Germans can afford this, most people around the globe are devastated when high electricity prices occur.[428]

We are killing people over solar panels and wind turbines that cause emissions to rise since solar and wind electrical generation plants need continuous, uninterruptable 24/7/365 backup from fossil fuels and nuclear. Renewables are unstable, intermittent sources of electricity.

Those two main reasons, unstable and intermittent, are why renewables need continual fossil fuel, or nuclear backup, and emissions are higher in countries that are heavily using renewables (Germany) versus France, which has some of the lowest emissions since they mainly use nuclear generated electricity.[429] The west is committing electricity suicide for no good reason, or is there?

CONCLUSION

Winston Churchill warned the west about the dangers of socialism leading to complete control of your daily and national lives.[430] Billions need energy leading to electricity; they don't need socialism, open borders, redistributed wealth, collectivism, New World Orders, or individual freedoms squashed.

If electricity-socialism becomes the norm that relies on unreliable, expensive, and undependable wind turbines and solar panels for electricity then society will collapse. Western environmentalists understand this, but they want energy-socialism since populations would then be subjugated to more government control, higher taxes, and parceling out electricity.

This has become modus operandi of western environmentalists in government, corporations, non-profits, academia, and some religious institutions. New York State led by Democratic Governor Andrew Cuomo expertly demonstrates what takes place when electricity-socialism is implemented. South Australia, Denmark, and Germany high electrical prices bear out this fact.

New York electricity polices under Governor Cuomo have taken an expensive turn based on the December 9, 2019 report from the New York Citizens Budget Commission (CBC) report titled, *Getting Greener: Cost-Effective Options for Achieving New York's Greenhouse Gas Goals.* This report is addressing the negative policy impacts from the New York Climate Leadership and Community Protection Act (CLCPA).[431]

The CBC is a non-partisan, nonprofit, civic-minded organization dedicated to "achieving constructive change in the finances and services of New York City and New York State government."[432] A noble mission, but attempting to go all-electric has severe consequences, and this report expertly points this out.

The CLCPA was mandated July 2019. Conservative costs to only use renewable electricity while dissolving fossil fuels is forty-seven billion dollars while continuing to use natural gas, and possibly nuclear generated electricity.[433]

The CBC report believes current electrical use will be the same as it is today (February 2020 when this was written) into 2040, "as New York moves to a path of decarbonizing heating and transportation in New York, the total electric demand will rise to 211,100 Gwh by 2040."[434]

To meet this electrical demand "nearly 94,000 Gwh of additional renewables (solar panels and wind turbines) will need to be added, a total that is roughly double the amount to be added from offshore wind (37,800 Gwh) and distributed solar (8,400 Gwh) now set by the CLCPA."[435]

Those energy requirements (Gwh) and power capacity needing installation (MW) would require "11,395 MW of residential solar, 16,117 MW of utility-scale solar, 18,457 MW of on-shore wind and 16,363 MW of off-shore wind to meet the increased load estimated by CBC."[436]

These estimates are low since renewables have a lower energy density than fossil fuels by fifty to seventy-five percent.[437] For New York solar requirements to be met:

> "Nearly twenty-seven square miles of residential roofs would have to be covered by over 364.6 million solar panels to meet the 11,395 MW estimate. For utility-scale solar each MW will cover 7 acres so 112,816 acres or 176 square miles will be needed to meet the 16,117 MW of utility scale solar output estimate."[438]

In the harsh winter months on the East Coast when the sun doesn't shine for days at a time, and the wind isn't blowing where will electricity be generated for heating homes and businesses?

Wind is tougher to calculate since it is inconsistent, but assume a "4.8 MW on-shore wind turbine would mean that over

3,845 on-shore wind turbines would be needed to meet the 18,457 MW output estimate."[439] But here is the largest problem, and the January 3-4, 2018 Winter Peak. With CBC Forecasted 2040 Capacity Resources to meet and/or exceed CLCPA goals revealed wind resources went to zero. For twenty-four out of forty-eight hours wind turbines were unable to deliver any electricity to New York.[440]

During this January 3-4, 2018 timeframe there was an electricity storage deficit of 13,545 MWh.[441] Next calculate the cost of the energy battery storage system using metrics from the National Renewable Energy Lab (NREL) Report: *2018 U.S. Utility-Scale Photovoltaics-Plus Energy Storage System Cost Benchmark*, and the cost for battery storage systems is forty-seven billion dollars.[442] This cost doesn't include solar panels or wind turbines, or the land and water space needed for them.

These are conservative estimates, and likely could double or triple in cost, and the value of the land, and amount of deep earth, exotic minerals needed for renewables and battery storage systems. Wind turbines and solar panels have every part that produces electricity results from petroleum derivatives

Extrapolate this out further, and imagine the United States producing electricity only with the wind and the sun. The U.S. consumes:

> "(Roughly depending on weather conditions) 3,911 billion kilowatt-hours annually or an average of eleven billion per day. The cost of storage to supply this amount of electricity for one day, when no other source was available would be $1.1 trillion, or $2.2 trillion at the current cost of Li-ion batteries."[443]

This estimate doesn't include low pressure weather systems during the winter when the sun isn't shining. High pressure weather systems during the summer when the wind isn't blowing. Or when there is an overabundance of both that overloads electrical grids throughout the U.S. Energy battery storage systems also need to be replaced every ten to twenty years, and different regions have different weather patterns.

The dispersion of the wind, sun, and electrical storage doesn't occur simultaneously, and would have to be planned for by elected officials, energy and electricity commissions, and electrical grids would need to be torn down, and rebuilt with a non-invented, smart grid technology.

An all-electrical world is complicated, and so far, not technologically possible, so why are we moving towards this path?

Because it has nothing to do with sustainable energy, climate change, or man-mad global warming. For the U.S. it means more votes for the Democratic Party, for Europe it means votes and power for Green parties, and European Union officials.

This isn't a conspiracy-theory laden conclusion, but the facts are clear: we are not technologically advanced enough now, or decades ahead to run economies, or daily lives on the wind, sun, and battery storage systems that have to be in place to make them feasible. The wind and the sun are not capable of providing lifesaving, continuous. uninterruptable electricity. Neither source of electricity can replace the derivatives from petroleum that account for every product in our daily lives.

Why has a child (Greta Thunberg) become the face of the global environmental movement – because she can be manipulated – and it is a no-win situation to be an adult, and disagree with her without looking mean, uncaring, or engaging in emotional child abuse.

That is why both authors of this book want her, and her parents to receive professional counseling, and care. We want the best for her, but that doesn't mean she knows anything about an all-of-the-above approach to energy, how to deliver electricity globally, or to the over 2 billion without it currently.

China is enslaving millions of Uighur Muslims, Iran is the leading sponsor of terrorism in the world, Russia uses their oil and natural gas reserves as a weapon, and India uses billions of pounds of coal per year, but environmentalists and left-leaning governments stay silent. Easier to chasten passive, western voters.

If we want to extend human creativity, and longevity then start with the previous paragraphs problems for environmental cleanliness, stop blaming the west so you (environmentalists) can garner more votes and taxpayer dollars, and work towards electrifying the globe.

When these "progressive petards (the left) has to live according to its own rules (for energy and electricity, and life itself) it will rue the loss of the civilization it destroyed."[444] We can do better, and we believe we will do better. Let's work towards electricity for the approximately two billion people without reliable electricity and leave the political power grabs behind.

CHAPTER SEVEN

CLIMATE ALARMISTS WHO BENEFIT FROM THE ALL-ELECTRIC NARRATIVE

By Todd Royal & Ronald Stein

INTRODUCTION

Roger Pielke is a contributor to *Forbes* magazine where he writes about science, technology, and gives unbiased research on energy, climate, and clean energy. His work highlights this fact. In 2018 the Intergovernmental Panel on Climate Change (IPCC) said:

> "Limiting global warming to 1.5 Celsius would require rapid, far-reaching and unprecedented changes in all aspects of society. Global net human-caused emissions (anthropogenic global warming) of carbon dioxide (CO_2) would need to fall by about 45 percent from 2010 levels by 2030, reaching 'net zero' around 2050."[445]

Mr. Pielke took this data a step further and showed how mathematically it is impossible to reach this goal of net zero emissions by 2050, or anytime in the near, or distant future. Mr. Pielke isn't a climate denier and believes "climate change poses a risk."[446] He also believes "aggressive mitigation and adaption policies make good sense."[447]

But we have a problem, and the climate alarmists want this to stop: *The British Petroleum (BP) Statistical Review of World Energy 2019* overwhelmingly highlights the world is consuming and using more fossil fuels and electricity. This growth will keep growing for decades ahead, because of China, India, and Africa. In 2018 the world consumed approximately 14,000 tons of oil equivalent (mtoe).[448]

Even more frightening to the climate alarmist crowd is:

> "From 2000 to 2008, consumption (fossil fuels) grew at about 2.2 percent per year and ranged from a drop of 1.4 percent in 2009 to an increase of 4.9 percent in 2004."[449]

This will keep growing since there are over two billion globally who are still demanding reliable electricity, and the products from petroleum derivatives.

Working off Pielke's math based off BP's projections here's why we must find a different path than zero-carbon societies, de-carbonization, or net-zero emissions.

Using only fossil fuels, as a basis of emissions (discounting livestock, land, buildings, burned cow dung and rotted wood, etc.), and understanding coal, especially the heavy usage in China and India, accounts for roughly forty percent of CO_2 emissions according to the Global Carbon Project. Let's understand the math behind a zero-carbon/carbon free society that only uses electricity for all societal needs and wants.

The IPCC wants a forty-five percent reduction in emissions by 2030, then 5,950 million of tons of oil equivalent (mtoe) of fossil fuels must be eliminated in ten years. (This chapter was written in February 2020). With consumption growing at a conservative estimate of 2.2 percent per year to this 2030 target then the world will add another 4,200 mtoe in 2030 from 2018.[450] To counter this fossil fuel growth the world will need to add over 10,000 mtoe of carbon-free electricity to hit the forty-five percent emission reduction target, and implement social adjustments to living with lesser numbers of products from petroleum derivatives.

However, there are a couple of problems with this solution. Clean electricity never accounts nor can provide for the over six thousand products that come from a barrel of crude oil. Every part in wind turbines and solar panels comes from the derivatives from petroleum, and if the world's largest economy – the United States – shut down everything that caused emissions, including human life, emissions will still grow, because of China, India, and Africa's energy growth.[451]

Literally, the U.S. could have 100 percent reductions in CO2 emissions, and global emissions will rise, because of China, India, and Africa and their prolific usage of coal.

Back to Pielke's math – if we are adding 10,000 mtoe – based on 2.2 percent per year growth then we need to begin adding "1,000 mtoe of carbon-free electricity every year over the next decade. The world added sixty-four mtoe of carbon-free electricity every year from 2010-2020 with 2018 being a record year at 114 mtoe."[452]

Therefore, every country that are United Nations signatories needs to begin adding nine to fifteen times the rate of solar panels and wind turbines if they want to meet clean electricity goals.

Plus, all new fossil fuel deployments need to stop immediately. China is doing the exact opposite by adding more coal-fired electrical generation volume than the entire "capacity of the European Union."[453] This increase in fossil fuels coincides with:

"Over the past decade fossil fuel consumption increased by about 150 mtoe. 2018's record of

114 mtoe of carbon-free energy was dwarfed by
an increase in fossil fuels of more than 275 mtoe."

The 2.2 percent growth threshold was larger than antici-
pated. The world is smashing fossil fuel consumption records.

Demand for fossil fuels needs to be immediately eliminated,
along with the social changes necessitated by the loss of all the
products made from petroleum derivatives to meet IPCC goals;
and electrical demand satisfied by fossil fuels (coal, oil, natural
gas) needs to be immediately eliminated.

If net-zero carbon societies according to the IPCC are to
be met by 2050 then the "deployment of a new nuclear plant
every day for the next thirty years, while retiring the equivalent
amount of coal and natural gas electrical generation every day,"
needs to start now.[454]

More importantly since the IPCC wants a 1.5-degree Celsius
reduction by 2030, more carbon-free electricity needs to be com-
missioned, and brought online, beginning today. The sticking
point is how do we convince the public that they need to start
adjusting to living without the six thousand products from petroleum derivatives.

If the IPCC wants to make electricity "rapid, far-reaching and
unprecedented changes in all aspects of society," a degree Celsius
reduction, and a 50 percent de-carbonization around the world
then demand for all petroleum derivative products must stop
now. Society needs to change their wants and stop demanding a
prosperous world that begins with fossil fuels.

But somehow, no one is doing the math about what needs
to be eliminated versus what needs to be planned for, approved,
built, tested, and deployed (average nuclear plant takes decades to
build) to meet any future goals in 2030, 2040, or 2050. Realistic
climate discussions never take place.[455]

Climate alarmists view books like Marc Morano's *The
Politically Incorrect Guide to Climate* Change, or Gregory

Wrightstone's *Inconvenient Facts: The science that Al Gore doesn't want you to know* as scandalous, climate-deniers who should be disregarded, harassed, bullied, physically intimidated, and even threatened with death.

Full disclosure: I voted twice for the Clinton/Gore ticket; the facts are what matters. The former Vice President Gore should be questioned about his climate views, and the amount of money he has made off these views. Both authors of this book are in favor of that taking place.

Nor are learned, careful thinkers such as Roger Pielke, Drs. Patrick Moore, Judith Curry, Jay Lehr, or Richard Linzer listened to, or taken seriously by global media outlets when it comes to CO2, climate change, man-made global warming, or serious discussions regarding energy and climate policy.

Meanwhile, environmentalists, led by a teenager (Greta Thunberg) can't decide if ocean currents are speeding up, slowing down, or unsure, because of global warming; and various research papers contradict one another.[456]

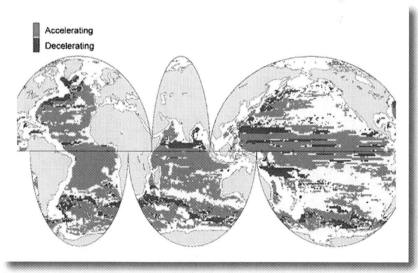

Global Ocean Circulation since the 1990s.

If Ocean's weren't confusing enough – now NASA – has imagined a "Little Ice Age" in the future, because of solar activity that rises and falls in 11-year cycles; and the newest one begins this year, 2020.[457] Climate change and high temperatures is parroted every day, and caused the British Broadcasting Company (BBC) "to partner with Greta Thunberg for (another) TV climate series."[458]

Without the existence of human beings or fossil fuels to blame for the previous five warming cycles that melted the ice from the previous five ice ages, we are left with a troublesome question.

Namely, how can the presence of humans and fossil fuels, for "0.00288[th] of a second" on the "24-hour clock," on the twelve percent of the earth's surface that is habitable land mass, have any influence, as compared to all the natural forces that have caused the five previous warming cycles and climate changes over the last four and a half billion years? [459]

The BBC, and Ms. Thunberg are contradicted by one of the greatest living, and recently deceased theoretical physicists, Dr. Freeman Dyson. Dr. Dyson worked at Princeton University, as a contemporary of Albert Einstein, and advised the U.S. government on a wide scope of scientific, and technical issues for decades. During the second term of the Obama Presidency he said:

> "The climate models used by alarmist scientists to predict global warming are getting worse (confirmed here),[460] not better; carbon dioxide does far more good than harm; and President Obama has backed the "wrong side" in the war on "climate change."[461]

Could Dr. Dyson be correct, and climate models are wrong? According to the National Snow & Ice Data Center (NSIDC), "Arctic Sea Ice is once again GROWING, with current 2020 levels exceeding eight out of the previous ten years."[462] Then it was announced in early February 2020 by the International Energy Association (IEA) that:

> "CO_2 emissions in 2019 outside advanced economies are growing in China, India, Africa, most of Asia, and Russia, but CO_2 emissions leveled off in advanced economies by switching from coal to natural gas, and higher nuclear power output."[463]

So why is the world environmental movement, the U.S. Democratic Party, Green-aligned European political parties, most large corporations, international foundations like the Ford Foundation and Rockefeller Foundation, and leading academic institutions worried about global CO_2 emissions?

Who stands to gain from fear and alarmism? The why of who benefits is the most fascinating part about the push to electrify economies, and societies with intermittent electricity using wind turbines and solar panels.

The above data shows from the previous pages there is nothing alarming, worrisome, or catastrophic about the environment. Ice caps are growing, polar bears are populating, the sunspots are cooling the earth, and CO_2 emissions are decreasing from west. Worry about emissions skyrocketing from China, India, and Africa for decades ahead if you want to worry about the environment.

Yet prophecies of climate-doom are all we hear from Tom Steyer, Michael Bloomberg, Naomi Klein, Bill McKibben, Bernie Sanders, Al Gore, and the entire western-led environmental movement.

Daily headlines scream environmental catastrophe, but if we want to concentrate on how China, India, and Africa are destroying their lands that is a worthy conversation. Grandstanding to gain votes, money, and power is what the climate alarmism movement wants more than anything.

Who are its leaders, and why they eschew science and debates to deliver unquestionable truths should leave anyone skeptical, as to their motives, reasons, and actions?

THE LEADERS OF THE CLIMATE ALARMISM MOVEMENT

If you examine the chart below it is interesting to note that most of the U.S. are still "net petroleum importers," with the U.S. being a net petroleum exporter.

U.S. states that vote to ban-fracking, embrace and deploy electricity from renewables over natural gas, coal, or nuclear are importing petroleum from countries, and areas like Russia, and the Middle East.[464]

Despite the U.S. becoming a net petroleum exporter, most regions are still net importers

Monthly total petroleum net trade (Jan 2010-Nov 2019) million barrels per day

The United States has been a net exporter of petroleum since Sep 2019

Midwest
East Coast
West Coast
Rocky Mountain
U.S. total
Gulf Coast

Source: U.S. Energy Information Administration, Petroleum Supply Monthly

If the U.S., and other western-aligned powers are going all-in for renewables then someone must benefit? Renewables only work with large taxpayer subsidies, feed-in-tariffs, loans, and government-guaranteed contract prices.[465]

There is no commercial value to consumers, businesses, or governments for the wind and sun for electricity unless all-of-the-above mechanisms are in place. Does Al Gore, Naomi Klein, Bill McKibben, and Senator Bernie Sanders care that NASA says the earth is greener today than it was twenty years ago – thanks to China and India's tree-planting programs?[466] Facts no longer seem to matter when the words global warming, or climate change are echoed.

This climate-industrial-complex has taken on an onerous, insidious form. They will lower human life from deploying expensive, unreliable electricity from the wind and sun. But this form of intermittent electricity provides none of the products from petroleum derivatives that are the cornerstone of developed countries.[467]

This path has been endorsed by scientific organizations that include the "American Geophysical Union, The American Association for the Advancement of Science, The American Meteorological Society, National Academies of Science, and the American Medical Association" to only name a few.[468]

Electricity cannot produce the over six thousand daily products that consist of products such as vaccinations, roads, bridges, water systems, and vanity products like makeup.[469] Electricity is nothing without energy that is abundant, reliable, affordable, scalable, flexible, and if you put nuclear into the mix then it is carbon-free.

Electricity is a stand-alone entity that needs coal, natural gas, nuclear, renewables (mainly the wind and sun from solar panels and wind turbines), biomass (wood, cow dung), or hydropower (dammed water) to produce anything.

So, knowing all this, who benefits from climate alarmism? The same people who claimed global cooling in the 1970s, are now claiming man-made (anthropogenic) global warming. This false narrative how the-world-is-going-to-end-nonsense has been going on for over 140 years.[470] *The New York* Time wrote about global cooling and scientific consensus in 1974-75.[471]

These climate alarmists will do anything to gain power, such as using children to promote their false environmental doctrines. British psychologists have "warned of the impact on children of apocalyptic discussions of climate change."[472] Another children's expert has said, "There is no doubt in my mind that they (children) are being emotionally impacted."[473]

Children are now suffering eco-anxiety from mythical, made-up environmental garbage. The same junk science has been promoted for decades, but now it is sinister, and evil since it involves children.

This isn't some *X-Files* television show with both authors of this book going back and forth, as Mulder and Sculley, looking for conspiracy theories, and gotcha-moments. We are agnostics on energy, climate change, and global warming that just want to show the facts.

Forty years ago, it was global cooling, today its global warming. What will it be tomorrow? If you don't believe me then research for yourself the previous paragraph's endnotes. Imbedded in those footnotes are direct quotes from the leading newspapers, and media publications in the world.

Ask yourself why Paul Ehrlich wrote *The Population Bomb* in 1968, and nothing came true from his book? Bill McKibben of Vermont wants clean electricity but hates carbon-free nuclear generated electricity. Why?

Mikhail Gorbachev, Ted Turner, Jacques Cousteau, Prince Phillip, and Henry Kissinger all believe in eliminating people, because the earth is running out of resources, overheating, and only people like them have the answers.[474] (This is a long endnote, but I challenge you, the reader, to read the entire source.)

Tom Steyer, the California billionaire environmentalist who made billions financing coal-fired power plants and has never once said he would divest his holdings. All Steyer wants is money and power. If there weren't gullible taxpayer dollars in it for him, he would slither off somewhere else looking for an easy buck.

Michael Bloomberg who is worth fifty-nine billion dollars, has a media and financial empire that uses gargantuan amounts of continuous, uninterruptable electricity to power his computer server farms moving information all over the world. At least Bloomberg created a great product (Bloomberg machine), and a world-class media company.

He receives his electricity from fossil fuels, and every day Bloomberg is using products that come from a barrel of crude oil. But he, and Steyer are supposedly green. No, they aren't. They are hypocritical, faux environmentalists, masquerading as environmental crusaders.

The newest, palaver-based, unwise environmentalists go by the name: Extinction Rebellion.[475] This is a corporate-owned protest group; also using fossil fuels and exotic minerals from human rights abusers every time they use their iPhone to text, or tweet the next location of their environmental protest.

They are paid activists, and care nothing about helping two billion people without electricity, or how the U.S. cut global emissions by using and advocating for natural gas-fired power plants for electricity. They, like Steyer and Bloomberg, are only in it for the money and power – nothing more, nothing less.

Not unlike celebrities such as Benedict Cumberbatch, Stephen Fry, Emma Thompson, Olivia Colman, Ellie Goulding, Tom Yorke, and Bob Geldof who have all promoted Extinction Rebellion.[476] None of these non-energy literate entertainers realize they are alive, because of the over six thousand products that come from a barrel of crude oil, i.e., the derivatives from petroleum. All these uneducated jesters are the best form of "useful idiots" imaginable.[477]

Extinction Rebellion, its corporate, and celebrity supporters, are narcissistic; neophytes used for others political gain. They want extinction of mankind the way Ehrlich does, and unhappy men like Steven Best, Norman Borlaug (father of the green movement), Albert Bartlett, Max Born, and Kenneth Boulding all use their God-given abilities and education to rule people in developed countries; instead of lifting up those in under developed countries from soul-destroying poverty for the betterment of mankind.

This form of corporate-government-environmentalism is meant to move people from suburbs, villages, farms, or wherever someone chooses to live in a free society into cities, and Soviet-styled apartment buildings.

Move a family from a single-family home into the city, and lives and properties are easier to control according to Joel Kotkin who is the Presidential Fellow in Urban Futures at Chapman University (California), and Executive Director for the Center for Opportunity Urbanism.[478]

The environmental movement needs to continue creating a global warming scare but has serious issues when probed deeper.

This takes on some ridiculous, fantastical movie where United Nations bureaucracies and corporate-backed Non-Governmental Organizations call on James Bond to herd the masses into cities and buildings to control lives and strip away the rights to defend their persons, and families.

Where are the global investigations about the climate-gate fiasco from 2010 when climate scientists colluded to manipulate climate data, and then lied about the colluded manipulation so they could garner larger budgets?[479] Build an extinction rebellion against faulty science, and eliminate taxpayer monies, then witness the entire environmental ideology crumble.

Steyer, Gore, Bloomberg, the U.S. Democratic Party, and European Union commissioners don't want to environmentally clean up China, India, and Africa, because that is hard work, and will take a lifetime without billions flowing into their pockets. A global contextual solution around fossil fuels, carbon-free nuclear, and private research dollars into building a better solar panel and wind turbine should be the environmental movement's answer.

But some nonsensical, global resource allocation model that only the special ones like Henry Kissinger understood is in the offing. We know the solutions: natural gas, nuclear, expanded coal fired-power plants for emerging countries, establishing petrochemical facilities for said countries to experience better food, clothing, infrastructure, and health care – those are the solutions.

Otherwise, an evil, global, environmental order led by the United Nations, or heirs to the Rockefeller fortune will lead to their profit via climate alarmism, and for the other ninety-nine percent of humanity – our standard of living is lowered, human progress is halted, and ingenuity takes a back seat so billionaires can have green virtue. Because, it was never about the climate, environmental health, or curing respiratory illness.

THE ENVIRONMENTAL MOVEMENT ISN'T ABOUT THE CLIMATE!

Currently, Americans, Europeans, multinational corporations, governments adhering to climate change policies, foundations, and NGOs are all in danger from the "forced electrification" movement.[480]

This insipid environmental cause wants to get rid of natural gas for home and businesses – (because it is never about the environment, only political and financial power) – and only garner intermittent electricity from solar panels and wind turbines.

Never mind that everywhere renewables are deployed emissions increase, while homes, and businesses powered by natural gas are responsible for approximately "one-third fewer greenhouse gas emissions."[481]

Daily lives are then entirely dependent on the electrical grid. No grid on planet earth can handle that amount of use, or daily load. It is untested, hasn't come close to being invented, and has zero sustainability.

Further, it is deeply troubling for reliable electricity globally that climate change has become the cause du jour for sensible, realistic energy policies. Since this chapter has unequivocally stated it isn't about the environment, expanding human prosperity, or the good of humanity, here is one basic question for the western-based, environmental cause.

- Is carbon dioxide (CO_2) an existential threat to planet earth?
- According to Dr. Patrick Moore, one of the founders of Greenpeace it isn't. CO_2 absorbs energy, but "the most important greenhouse gas is not carbon dioxide; it is water vapor."[482]
- Water is the fundamental component to the earth's climate, but if CO_2 doubles, triples, or goes higher in the atmosphere, doesn't that effect global air temperature?

Climatologists will refer to "equilibrium climate sensitivity – which is the eventual rise in air temperature due to a doubling of carbon dioxide."[483] The problem is global media outlets, governments, and environmentalists that have said ad nauseum the climate is warming, but "climate sensitivity have decreased substantially, based on measurements of the climate system."[484]

In the early to mid-2000s estimates had CO_2 causing temperatures to rise between three and six degrees Celsius, since 2010 estimates have climate sensitivity at three degrees Celsius or less. After 2015 "independent assessments have placed the sensitivity at about one degree Celsius or 1.8-degree Fahrenheit."[485]

Climate models and theories both state: "the biggest effect of carbon dioxide on air temperatures should lie in the upper tropical troposphere. The troposphere is the layer of the atmosphere where all weather resides."[486]

For over forty years the warming has been smaller than predicted, and "daily maximum air temperatures have not changed substantially over the last eighty years, and before that, maximum air temperatures were much higher during the Dust Bowl of the 1930s."[487]

This means CO2 has less impact than previously studied or thought. This is likely why climate models are inaccurate, and all predictions of the last fifty years, or even 140 years about the climate are wrong. The climate isn't a number that can simply be averaged into a discernable equation.

Similar to how only using solar panels and wind turbines cannot solve electrical generation demands, or building new electrical grids needed to handle their intermittent nature. The earth, like energy policies, is dynamic, evolving, and always changing.

Now the biggest question of all: is global warming real, has mankind caused the earth to warm irreparably? The earth has warmed 0.6 degrees Celsius since the late 1800s – post Industrial Revolution.

But humans located in dense, urban societies cause warming in their vicinity. The Washington D.C. Metropolitan area as an example cause – the "urban heat island" effect – is warmer than the surrounding countryside and lower population counties and states. The same could be said for Beijing, Budapest, London, Rio de Janeiro, name the large urban center, and the urban heat island effect is real.

This urban heat phenomenon holds true for any densely populated urban areas over wide-open space, or single-family neighborhoods. It's why saying "this year is the warmest in recorded history" is horribly misleading.

You take minimum daily air temperatures, which have increased due to the urban heat phenomenon, average those numbers together from weather stations showing hotter numbers, and you have warming that isn't due to CO2.

It is also why urban areas have intensified floods, drought

frequencies, wildfires, and other natural calamities over less populated areas. Weather catastrophes have more to do with water usage and land use management than CO2.

Then is CO2 the main driver of human catastrophe in the minds and policy recommendations of environmentalists? The answer is no:

> "A warmer climate and more carbon dioxide are a net benefit to life since more people die from cold than heat, (because CO2 causes) longer growing seasons. Further since CO2 is plant food, under higher carbon dioxide concentrations, virtually all plants grow faster and are more efficient in using water."[488]

It is beyond energy-wise comprehension, and folly for global cities – say Los Angeles, London, Tokyo, or Paris – to embrace, and advocate for Green New Deals.[489] Spending trillions on CO2 emission reductions, or stabilizing the climate by advocating for only electricity from the sun and the wind is madness: when the past few pages have shown CO2 is a small instigator in climates changing.

Seemingly it is about the United Nations taking wealth from first world nations and giving them to third world nations with the UN being the one doing the redistributing. This does nothing to clean up China, India, Africa, Asia, or deliver reliable electricity to billions.

Climate change is the cause-celebre for power and control when a natural gas-fired; coal-fired, or nuclear power plant is what these billions need to deliver to them reliable, stable, and life-giving continuous, uninterruptable electricity.

The Green New Deal is a deeply misguided power grab when you realize, "since 2005 the U.S. has cut back on greenhouse gas emissions by about thirteen percent," with zero effect on the earth's climate."[490] The U.S. also "cut its emissions more than any other country in the world in 2019," but "the gains were offset by the growing emissions from the rest of the world," according to the International Energy Agency's 2019 report on CO2 emissions.[491]

The U.S. could literally shut down: every living thing ceases to exist, everything is destroyed, the U.S. is back to its natural state, but emissions would still rise globally.[492] The Green New Deal and its backers only want votes, power, and money. They do not care about reliable electricity, or global health leading to longer lives.

Green New Deals that want to cut CO2 emissions eighty percent by 2050 are engaging in unrealistic math and engineering, as this chapter discussed in the introduction. Economic progress is "primarily dependent on its energy sector," and "energy is the backbone of any developing country."[493]

If a country, or continent does not have abundant, reliable, scalable, affordable, flexible and electricity then you do not have an economy, or improved living standards. That is why climate alarmists want Green New Deals, global warming, climate change, and renewables, to shut down progress, and human equality.

Today's environmental movement has never been about the climate, and likely never will. Reliable electricity equals innovation, infrastructure, and life itself. The environmental movement's goals are the exact opposite.

THE WORLD IS BETTER WITHOUT CLIMATE ALARMISTS

"Billions will die," that's the conclusion Extinction Rebellion gave to British television in October 2019. No one questioned them or asked what scientific proof they gave for saying billions are going to die. To give perspective – the deadliest war in the history of mankind – World War II, saw fifty to seventy million deaths occur.[494]

In response to this climate alarmist manipulation of children, Lauren Jeffrey's, a seventeen-year-old high school student in a city fifty miles northwest of London posted a YouTube.com video eloquently saying to Extinction Rebellion:

> "As important as your cause is, your persistent exaggeration of the facts has the potential to do more harm than good to the scientific credibility of your cause as well as to the psychological well-being of my generation."[495]

There is zero scientific evidence we are going extinct, as a species, or human race.[496] Nevertheless, scientific evidence doesn't stop the environmental movement, and climate alarmists backed by a daily barrage of doom and gloom from global media outlets.

Bill McKibben believes climate change caused fires in Australia that have now made koalas "functionally extinct."[497] Media firm *Vice* recently stated, "the collapse of civilization may have already begun."[498]

Former New York bartender, and now U.S. Congresswoman, Alexandria Ocasio-Cortez, the author, and proponent of the American Green New Deal claims, "The world is going to end in 12 years if we don't address climate change."[499]

Where is the evidence for these claims? None, when the CO2 arguments have serious flaws that should be debated and understood. The definitive report on global warming, and climate change comes from the United Nation's *Intergovernmental Panel on Climate Change*.

Nowhere in these exhaustively researched documents do you find references to billions dying, children being extinguished, mass migration over climate change, sea levels rising to unhealthy levels, fire-causing global warming, crop failure, famine, or wars engulfing the world over man cooking the earth's temperature to unsustainable levels by using fossil fuels (oil, petroleum, natural gas, and coal).[500]

More importantly, there are still over six thousand products that come from a barrel of crude oil. Global emissions will rise over the use of fossil fuels from China, India, and Africa, and 2 billion people without reliable electricity will do whatever it takes to achieve their electrification goals.

They will never listen to faux-environmentalists, masquerading as climate-crusaders. To them the west is a bunch of lying, immature, slovenly-base, climate-change-imbeciles.

What is known, factual, and scientifically sound are the following researched, and sourced points:

> There is a 99.7 percent decline in deaths from natural disasters that peaked at 3.7 million in 1931 to 11,000 in 2018.[501]

The United Nations Food and Agriculture Organization (FAO) forecast:

> "Crop yields increasing thirty percent by 2050, and the poorest parts of the world, like sub-Saharan Africa, will see increases of eighty to ninety percent. Humans produce enough food for ten billion people."[502]

Twenty five percent more food than we need since there are approximately 7.6 billion people on the earth, and our food production figures are projected to rise dramatically in coming decades. Thirty different climate models "found that yields would decline by six percent for every one-degree Celsius increase in temperature."[503] This is likely why the IPCC doesn't give dire climate alarmist predictions.

Climate alarmists are also wrong that climate change will decimate economic prosperity. The IPCC "anticipates climate change will have a modest impact on economic growth," and by "2100 IPCC projects the global economy will be 300 to 500 percent larger than it is today."[504] This chapter was written in February 2020.

The IPCC expects global warming of 2.5 to four degrees Celsius.[505] Renowned Yale University economist William Nordhaus predicts the same temperature rise.[506] Both agree this would reduce global gross domestic product (GDP) by "Two percent and five percent over that same period."[507] That is not a calamitous number.

With regards to climate change causing fires – yes, the climate is always changing, and drier temperatures increase fire risk – but according to Richard Thornton of the Bushfire and Natural Hazards Cooperative Research Centre in Australia, "Climate change is playing its role here (The deadly Australian fires from late 2019), but it's not the cause of these fires."[508]

Australia's indigenous people have a solution for the country's bushfires that they have been using for roughly 50,000 years. "Aboriginal people had a deep knowledge of the land. They can feel the grass, and know if it would burn well; they knew what types of fires to burn for what types of land, how long to burn, and how frequently." said historian Bill Gammage, an emeritus professor at Australian National University who studies Australian and Aboriginal history.[509]

U.S. scientists modeled thirty-seven different regions, and "found humans may not only influence fire regimes but their presence can actually override, or swamp out, the effects of climate."

The same scientists also found ten variables that influence, or cause fires, however, "None were as significant as the anthropogenic variables." Meaning, fires aren't caused mainly by global warming, or climate change, but such mundane things as "building homes near, and managing fires and wood fuel growth within, the forest."[510]

Good for climate scientists beginning to push back against the lies, falsehoods, horrible energy polices, and outright exaggerations. Australian climate scientist, Tom Wigley, who started working on climate science in 1975, and created "the first climate models (MAGICC) in 1987, it remains one of the main climate models used today."[511] Mr. Wigley remarked in 2019:

> "It really bothers me (the climate change threatens civilization narrative), because it's wrong. All these young people have been misinformed. And partly it's Greta Thunberg's fault. Not deliberately. But she's wrong."[512]

The most important part about our climate, and our world is that the world is dramatically improving. The climate alarmists continue getting these facts wrong.

Average global life expectancy doubled (conservative estimates) since 1900 and is approximately above seventy years of age per person.[513]   Global literacy has increased leading to better family planning, increased healthcare access, and greater human rights; and equality for girls and women.[514]

While forced child labor in any form is awful, and condemned on these pages, it is decreasing.[515] We have greater peace than almost any time in recorded, world history.[516]

Moreover, global incomes correlating with GDP, "over the past thirty years, global *per capita* income has almost doubled."[517] This has led to incredible poverty reductions in countries such as India, China, most of Asia, and Mexico. Even open defecation practices have gone from thirty to fifteen percent globally from 1990-2015.[518]

During the same 1990-2015-time period, "2.6 billion people gained access to improved water sources, bringing the global share up to ninety-one percent."[519] The improved environment has caused respiratory deaths from air pollution to "decline substantially," globally, and in low-income countries.[520]

Wealthy, first world nations, are witnessing increased forestation, and preservation since higher food yields are allowing less land use for farming and livestock.[521]

This is only a snapshot for how well our world is doing in all aspects of progressive human longevity, and growth. To understand a fuller picture please read Ronald Bailey's book, *The End of Doom: Environmental Renewal in the Twenty-first Century*, and Gregg Easterbrook's book, *It's Better Than It Looks: Reasons for Optimism in an Age of Fear.*

CONCLUSION

Scotland taxpayers and businesses have paid out more than 650 million Sterling Pounds to not generate electricity (called constraint payments when demand falls, or the wind is blowing to hard) to Scottish electrical users.[522]

British Petroleum (BP) said in February 2020 that it would cut emissions from its entire global operations to zero by 2050, and cut consumers burning fuels.[523] BP will offset some of the emissions from the oil it produces with trees or carbon capture, but bringing that to zero carbon emissions means it will be producing far more low-carbon energy, and far less oil and gas in 2050. This seems more of a catch-all ploy to sooth environmental zealots, and media outlets than anything resembling reality from an energy company.

The U.S. state of Virginia's house and senate passed legislation to phase out coal, and go 100 percent renewable electricity between 2045-2050.[524] Neither U.S. Democratically controlled governmental body declared how they would rid the state of the over six thousand products from society that come from a barrel of crude oil, or how they would overcome the unstable, unreliable, intermittent nature of wind turbines and solar panels.

Shell Oil and EDP Renovaveis SA have agreed to a joint wind turbine farm to generate electricity for the ratepayers of Massachusetts that both companies are claiming will supply electricity at a "record-low price."[525]

Unless both firms can overcome the Scottish example in the first paragraph of this section, or they are using a wind turbine that is currently unheard of, the facts state the exact opposite will take place with electrical prices (they will skyrocket like they've done in Germany and Australia) to ratepayers in Massachusetts from a wind farm, built with the products from petroleum derivatives, off their Atlantic coastline.

BlackRock, the world's largest investment management firm, has decided through its green-oriented funds they will no longer invest in Alberta, Canada oil sands projects; or coal.[526]

They believe it is destructive without disclosing what will assist giving two billion people reliable electricity that coal provides. Nor will BlackRock condemn China, or India for their dramatic uptick in building, and using coal-fired electrical plants.

Other financial giants, and BlackRock who are climate sensitive seemingly have no idea that climate models are not showing warming in the Arctic.[527] Without financing Alberta oil sands projects, blue collar workers, and every day people like teachers, and truck drivers are hurt so wealthy financiers can feel good about themselves at their next cocktail party.

Australia has been phrased, "Renewables-World down under," after most of Australia, and specifically South Australia have almost had complete electrical grid meltdowns, and blackouts based on over-relying on wind turbine farms.[528] Australia like Germany will have imminent collapse of their grid, societal delivery of electricity, and put their countries at risk of invasion if they continue their transition to renewables (wind turbines and solar panels) for intermittent electricity.

Climate has become environmentalists' religion. A dogmatic, ideological belief "that is impervious to contradiction by logic, evidence or experience, and cultivate a moral superiority towards unbelievers."[529] That describes the western-based, US-led, and European affirmed environmental movement.

Traditional environmentalism that worked across ideological, political party lines to protect endangered species, and advocate for clean air, water, and the earth itself – now wants to ban fracking, rid the world of fossil fuels, and decarbonize/de-industrialize the entire world – using unrealistic, Green New Deals.[530]

What environmentalists want now by electrifying the world is complete control over how people eat, drink, work, how many children they have, what you drive, and where you live. The radical activists like Naomi Klein and Tom Steyer want to ban, "eating meat, driving SUVs, flying and using plastic straws."[531] These are mortal sins to nihilistic radicals, masked as environmentalists.

The green purists carve the world into worshipping the earth, and climate deniers. Just electric from the wind and sun is the only answer, everything else is shunned, discredited, and destroyed. Mock the denier of climate change. Shun anyone who doesn't buy the energy policy that only uses electricity.

Moderates no longer exist within the environmental movement in classrooms, western governments, newsrooms, foundations, large multinational corporations, non-governmental organizations, and non-profits.

When zero-carbon nuclear electricity, and abundant, clean, efficient, and flexible natural gas are no longer options for electrical energy efficiency, and use in homes, transportation, and electricity then climate alarmists will ruin the western world.

The warning we give is cleaner air, and declining CO2 emissions isn't the goal it once was when U.S. and European lakes and rivers were burning, and the air was as dirty as it currently is in Beijing. Now entire economies, and abundant, fruitful lives need to be destroyed for a Green New Deal.

We urgently and with necessity ask you, the reader, to consider the environmental, climate alarmist, who are nihilists, and only want to profit, and rule with an iron fist all in the name of the environment.

What this chapter has shown – it has nothing to do with the environment – and everything to do with votes, political power, a return to feudalism where the earth-worshippers rule (Gore, Steyer, Sanders, McKibben, Klein, et al), and money.

Just Green Electricity is the title of this book with a sub-title *Helping Citizens Understand a World without Fossil Fuels* and discusses positive small steps that can be achieved where all prosper, and electricity is delivered globally. But if these national, socialist-environmentalists like Al Gore and Tom Steyer are not defeated and brought down to the rubbish-bin of history we will enter a dark time for energy policies, the west, and global security.

RONALD STEIN

Founder and Ambassador for Energy & Infrastructure
PTS Advance

Ronald Stein is the co-author of the newly released book, "Energy Made Easy," and an internationally published columnist. As founder of PTS Advance (Principal Technical Services – 1995), Ronald Stein has developed one of the most successful and innovative family owned professional services firms in California. Known as the leader in delivering staffing solutions to the twelve major oil refineries in the state, the business has since transformed to support a range of staffing, consulting, project services and business process outsourcing solutions to the wider Energy & Infrastructure, and Life Sciences industries.

Over the last decade, Ron has become the private business spokesperson for the energy and infrastructure industries. He is an energy expert who writes frequently about all aspects of energy and economics and is an energy policy advisor for The Heartland Institute. Through his hundreds of published Op Ed articles, he provides an education for citizens as to what and why the energy infrastructures are the primary infrastructures that truly drives the worlds' economies.

Ron is an energy agnostic who only wants to share the facts for the public to be further energy literate, through his books, Op Ed articles, radio and TV interviews, and in-person presentations, to provide in-depth discussions, and explanations on many energy related subjects.

TODD ROYAL

Independent public policy consultant in Los Angeles focusing on the geopolitical implications of energy

Todd Royal is an internationally published columnist and author with a growing consulting practice based in Los Angeles, California. His publication that is in the U.S. Library of Congress titled: "Hydraulic Fracturing and the Revitalization of the American Economy," helped launch his writing and consulting career. Todd took his time from working at Duke University on value chain analysis and using SWOT (strengths, weaknesses, opportunities, and threats) economic research into the $445 billion dollar global furniture industry, and current African aid programs administered by U.S. AID to complete groundbreaking research on how energy and economic recovery are intertwined. Todd is currently working for Duke using supply and value chain analysis on energy battery storage systems.

After graduating from Pepperdine University's School of Public Policy with his Master's in Public Policy (M.P.P.) with highest honors in International Relations and State and Local Government, Todd began writing for OilPrice.com, The National Interest, Asia Times, USA Today, Yahoo Finance, Business Insiders, ModernDiplomacy.com, SeekingAlpha.com, EurasiaReview.com, and the American Society of Civil Engineers on energy on energy, foreign policy, national security, and California politics.

But Todd's focus began shifting heavily towards energy and how energy is used as a weapon by Russia, China and Iran every bit as much as their nuclear arsenals. His work on energy, foreign policy and national security has gotten the attention from industry leaders, energy organizations and State and Federal officials.

ENDNOTES

Chapter Notes:
(Due to the dynamic nature of the Internet, the location of some of the more than 500 items cited in this work—and accessed at the time of writing—may change as menus, homepages, and files are reorganized.)

INTRODUCTION: *JUST GREEN ELECTRICITY*

1 *British Petroleum Statistical Review of World Energy 2019*. Release date, 11 June 2019. www.bp.com, https://www.bp.com/en/global/corporate/news-and-insights/press-releases/bp-statistical-review-of-world-energy-2019.html
2 Romel, Valentina, Reed, John, "The Asian century is set to begin," www.FT.com (Financial Times), March 25, 2019. https://www.ft.com/content/520cb6f6-2958-11e9-a5ab-ff8ef2b976c7
3 Ridley, Matt, "We've just had the best decade in human history. Seriously, Little of this made the news, because good news is no news," www.Spectator.Co.Uk, January 1, 2020. https://www.spectator.co.uk/2019/12/weve-just-had-the-best-decade-in-human-history-seriously/?fbclid=IwAR2SZKD9rAbNr7n1OamrX6A5Gn2Ei5FOCa6UTHRuoVvM-iAYUwkwEfOuHZQ
4 U.S. Department of Energy, "Electric Power," www.energy.gov, page accessed January 2, 2020. https://www.energy.gov/science-innovation/electric-power
5 United Nations (UN), Economic and Social Council > Commission for Social Development > Fifty-Seventh Session, 6th & 7th Meetings

(AM & PM), "Impact of Natural Disasters Increasingly Affecting Those Most Vulnerable, Speakers Say as Commission for Social Development Continues Session," www.UN.org, February 13, 2019. https://www.un.org/press/en/2019/soc4876.doc.htm

6 Beurden van Ben, "Two billion people do not have access to reliable electricity: this must change," www.Linkedin.com, October 18, 2018. https://www.linkedin.com/pulse/two-billion-people-do-hav e-access-reliable-must-ben-van-beurden/

7 Easterbrook, Gregg, *It's Better Than It Looks: Reasons for Optimism in an Age of Fear*, (Hatchette Book Group, New York, NY), February 20, 2018. Entire book to back up human progress and people living longer claim. https://www.amazon.com/Its-Better-Than-Looks-Optimism/ dp/161039741X

8 Mills, Mark P., "Inconvenient Energy Realities: The Math behind 'The New Energy Economy: An Exercise in Magical Thinking,'" www.Economics21.org, July 1, 2019. https://economics21.org/ inconvenient-realities-new-energy-economy

9 Menton, Francis, "Contrast Of Climate And Energy Policies, And Economic Results, In The U.S. And Germany," www. ManhattanContrarian.com, December 6, 2019. https://www. manhattancontrarian.com/blog/2019-12-6-contrast-of-climate-and -energy-policies-and-economic-results-in-the-us-and-germany

10 Axelrod, Tal, "Merkel vows Germany will do 'everything humanly possible' to fight climate change," www.TheHill.com, December 31, 2019. https://thehill.com/policy/energy-environment/476416-merke l-vows-germany-will-do-everything-humanly-possible-to-fight

11 Ranken Energy, "Products made from petroleum: With over 6,000 products and counting, petroleum continues to be a crucial requirement for all consumers," www.Ranken-Energy.com, page accessed January 2, 2020. https://www.ranken-energy.com/index. php/products-made-from-petroleum/

12 Medium Corporation Twitter Account, "Climate Change and the Ten Warning Signs for Cults," www.Medium.com, February 23, 2019. https://medium.com/@hwater84/climate-change-and-the-te n-warning-signs-for-cults-56c181db82c1

13 U.S. Energy Information Administration, Independent Statistics & Analysis, Today In Energy, "In 2018, the United States consumed more energy than ever before," www.EIA.gov, December 23, 2019. https://www.eia.gov/todayinenergy/detail.php?id=42335

14 Hao, Feng, Baxter, Tom, "China's coal consumption on the rise," www. ChinaDialogue.net, January 3, 2019. https://www.chinadialogue.net/ article/show/single/en/11107-China-s-coal-consumption-on-the-rise

15 ESI Africa Edition 5, "Low emission tech can save coal's dominance," www.Esi-Africa.com, December 4, 2018. https://www.esi-africa. com/low-emission-tech-can-save-coals-dominance/

16 Center For Strategic & International Studies (CSIS), China Power Project, China Power, "How is China's energy footprint changing?" www.ChinaPower.CSIS.org, 2019. Page accessed on January 2, 2020. https://chinapower.csis.org/energy-footprint/

17 Varadhan, Sudarshan, "Coal to be India's energy mainstay for 30 years – NITI Aayog report," www.In.Reuters.com, May 15, 2017. https://in.reuters.com/article/india-coal-energy/coal-to-be-indias-en ergy-mainstay-for-next-30-years-niti-aayog-report-idINKCN18B1XE

18 Society of Environmental Toxicology and Chemistry, "It's a small (coal-polluted) world, after all," www.Phys.org, December 20, 2019. https://phys.org/news/2019-12-small-coal-polluted-world.html

19 Tverberg, Gail, "How Renewable Energy Models Can Produce Misleading Indications," www.OurFiniteWorld.com, October 24, 2019. https://ourfiniteworld.com/2019/10/24/how-renewable-energ y-models-can-produce-misleading-indications/

20 Hartnett White, Kathleen, "Renewables are incapable of replacing hydrocarbons at scale," www.TheHill.com, March 30, 2016. https:// thehill.com/blogs/pundits-blog/energy-environment/274645-renew ables-are-incapable-of-replacing-hydrocarbons-at

21 U.S. Energy Information Administration, Independent Statistics & Analysis, "Oil: crude and petroleum products explained," www.EIA.gov, Page last updated by EIA May 23, 2019. Page accessed on January 2, 2020. https://www.eia.gov/energyexplained/ oil-and-petroleum-products/

22 Travelweek Group, "Exactly how many planes are there in the world today?" www.TravelWeek.ca, February 17, 2017. https://www. travelweek.ca/news/exactly-many-planes-world-today/

23 Slav, Irina, "Diesel Demand Is Set To Soar," www.OilPrice.com, September 16, 2018. https://oilprice.com/Energy/Energy-General/ Diesel-Demand-Is-Set-To-Soar.html

24 Smil, Vaclav, *Energy and Civilization A History*, (MIT Press, Cambridge, MA, London, England), Chapter 5, Section on Renewable Energies, Pages 284-9, further see Chapter 6 Fossil-Fueled Civilization,

Pages 295-384. November 13, 2018. https://www.amazon. com/Energy-Civilization-History-MIT-Press/dp/0262536161/ ref=sr_1_1?crid=3A3B2GERAANJ2&keywords=energy+and+ civilization+a+history&qid=1578087532&sprefix=energy +and+civ%2Caps%2C208&sr=8-1

25 Stein, Ronald, Royal, Todd, *Energy Made Easy: Helping Citizens Become Energy Literate*, (Xlibris US, Bloomington, IN.), Chapter 9: Requirements of a Carbon-Free Society, www.Xlibris.com, August 12, 2019. https://www.amazon.com/Energy-Made-Easy-Citizens-Energy-Literate/dp/1796049840/ref=tmm_hrd_swatch_0? encoding=UTF8&qid=1577763460&sr=1-3

26 U.S. Energy Information Administration, Independent Statistics & Analysis, Analysis & Projections, *International Energy Outlook 2019*, (U.S. Department of Energy, Washington, D.C.), www.EIA. gov, Release date: September 24, 2019. Next release date: September 2020. https://www.eia.gov/outlooks/ieo/

27 Rogers, Norman, "Green Energy Studies: Consulting or Advertising?" www.AmericanThinker.com, 27 November 2019. https://www. americanthinker.com/articles/2019/11/green_energy_studies_ consulting_or_advertising.html

28 Bailey, Ronald, "Power U.S. Using 100 Percent Renewable Energy is a Total Fantasy: New Research debunks a study claiming there's a low-cost way to power America using wind, solar and hydropower," www. Reason.com, June 21, 2017. https://reason.com/blog/2017/06/21/ powering-us-using-100-percent-renewable

29 Cohen, Ariel, "New Decade, Same Risks For Oil And Gas Markets," www.Forbes-com.cdn.ampproject.org, December 23, 2019. https://www-forbes-com.cdn.ampproject.org/c/s/www.forbes. com/sites/arielcohen/2019/12/23/new-decade-same-risks-for-oi l-and-gas-markets/amp/

30 Bailey, Ronald, "How Much Will the Green New Deal Cost? Climate change is the excuse; radically remaking the American economy is the aim," www.Reason.com, February 7, 2019. https://reason.com/ blog/2019/02/07/green-new-deal-democratic-socialism-by-o

31 Watson, Frank, "COP25: 285 companies commit to science-based emissions targets," www.SPGlobal.com, December 4, 2019. https://www.spglobal.com/platts/en/market-insights/latest-news/ electric-power/120419-cop25-285-companies-commit-to-sc ience-based-emissions-targets

32 Perry, Mark J., "50 years of failed doomsday, eco-poclyptic predictions; the so-called 'experts' are 0-50," www.AEI.org, September 23, 2019. https://www.aei.org/carpe-diem/50-years-of-failed-doomsday-eco-pocalyptic-predictions-the-so-called-experts-are-0-50/

33 Moore, Patrick via PragerU and YouTube.com, "What They Haven't Told You about Climate Change," www.PragerU.com, July 27, 2015. https://www.youtube.com/watch?v=RkdbSxyXftc

34 I & I Editorial Board, "Extinction Rebellion Founder Confirms That Global Warming Is Voodoo Science," www.IssuesInsights.com, December 30, 2019. https://issuesinsights.com/2019/12/30/extinction-rebellion-founder-confirms-that-global-warming-is-voodoo-science/

35 Shellenberger, Michael, "Why Everything They Say About The Amazon, Including That It's The 'Lungs of the World,' Is Wrong," www.Forbes.com, August 26, 2019. https://www.forbes.com/sites/michaelshellenberger/2019/08/26/why-everything-they-say-about-the-amazon-including-that-its-the-lungs-of-the-world-is-wrong/#34d651d45bde

36 Shellenberger, Michael, "Why Everything They Say About California Fires – Including That Climate Matters Most – Is Wrong," www.Forbes.com, November 4, 2019. https://www.forbes.com/sites/michaelshellenberger/2019/11/04/why-everything-they-say-about-california-fires--including-that-climate-matters-most--is-wrong/#536822384cb6

37 Shellenberger, Michael, "Why Apocalyptic Claims About Climate Change Are Wrong," www.Forbes.com, November 25, 2019. https://www.forbes.com/sites/michaelshellenberger/2019/11/25/why-everything-they-say-about-climate-change-is-wrong/#5aa0b92512d6

38 Admin, The Art of Annihilation, "The Manufacturing of Greta Thunberg – for Consent: The Political Economy of the Non-Profit Industrial Complex," www.theartofannihilation.com, January 17, 2019. http://www.theartofannihilation.com/the-manufacturing-of-greta-thunberg-for-consent-the-political-economy-of-the-non-profit-industrial-complex/

39 University of Copenhagen, Denmark research team, "Study suggests global impact of obesity may be extra 700MT/y CO2eq: about 1.6% of worldwide GHG emissions," www.GreenCarCongress.com, December 23, 2019. https://www.greencarcongress.com/2019/12/20191223-obesity.html

40 Lomborg, Bjorn, "How Climate Policies Hurt The Poor," www.Project-Syndicate.org, September 26, 2019. https://www.project-syndicate.

org/commentary/governments-must-reduce-poverty-not-emissions-by-bjorn-lomborg-2019-09

41 Thomas, Cal, "Greta Thunberg's message of climate doom misses the mark," www.WashingtonTimes.com, September 25, 2019. https://www.washingtontimes.com/news/2019/sep/25/greta-thunbergs-message-of-climate-doom-misses-the/

42 Harsanyi, David, "Greta Thunberg Is the Perfect Hero for an Unserious Time," www.DailySignal.com, December 13, 2019. https://www.dailysignal.com/2019/12/13/greta-thunberg-is-the-perfect-hero-for-an-unserious-time/?utm source=rss&utm medium=rss&utm campaign=greta-thunberg-is-the-perfect-hero-for-an-unserious-time?utm source=TDS Email&utm medium=email&utm campaign=MorningBell&mkt tok=eyJpIjoiTW1VeU1qRTFNR1ZpTTJVMSIsInQiOiJxYnhyWWxpSnlaMFhIQkhEdDV3VVBzaGxaV2FTWENJYk5XYkp4ZWVyTRcLlwvczV6bEJMVUJjT09yNW9od3k1MzlPKzJ4V05EOUNGejFUR3ZFd2JnWGtcL0FUdVwva3VPSlVxdm1wTU9SV0lnNTBwS2k3TCtuNklLZkRwXC80RWNOd3lUeElFIn0%3D

43 Parke, Phoebe, "Why are 600 million Africans still without power?" www.CNN.com, April 1, 2016. https://www.cnn.com/2016/04/01/africa/africa-state-of-electricity-feat/index.html

44 Hubbard, Ben, Karasz, Palko, Reed, Stanley, "Two Major Saudi Oil Installations Hit by Drone Strike, and U.S. Blames Iran," www.NYTimes.com, September 14, 2019. https://www.nytimes.com/2019/09/14/world/middleeast/saudi-arabia-refineries-drone-attack.html

45 Malley, Robert, "The Unwanted Wars: Why the Middle East Is More Combustible Than Ever," www.ForeignAffairs.com, November/December 2019. https://www.foreignaffairs.com/articles/middle-east/2019-10-02/unwanted-wars?utm medium=newsletters&utm source=fatoday&utm content=20200101&utm campaign=FA%20Today%20010120%20ISIS%27s%20New%20Caliph%2C%20Remembering%20Michael%20Howard%2C%20Volatility%20in%20the%20Middle%20East&utm term=FA%20Today%20-%20112017

46 Alexander, David, "U.S. envoy says Venezuela oil production dropping steadily," www.Reuters.com, March 15, 2019. https://www.reuters.com/article/us-venezuela-politics-usa-abrams/u-s-envoy-says-venezuela-oil-production-dropping-steadily-idUSKCN1QW2HM

47   Cohen. Ibid. 2019.
48   Cohen, Ariel, "Making History: U.S. Exports More Petroleum Than It Imports In September and October (2019)," www.Forbes.com, November 26, 2019. https://www.forbes.com/sites/arielcohen/2019/11/26/making-history-us-exports-more-petroleum-than-it-imports-in-september-and-october/#58b8f4c05f3b
49   U.S. Energy Information Administration, Independent Statistics & Analysis, Analysis & Projections, "Short-Term Energy Outlook," www.EIA.gov, Release Date: December 10, 2019. Next Release Date: January 14, 2020. https://www.eia.gov/outlooks/steo/report/index.php
50   Russell Mead, Walter, "How American Fracking Changes the World," www.WSJ.com, November 26, 2018. https://www.wsj.com/articles/how-american-fracking-changes-the-world-1543276935
51   Shiryaevskaya, Anna, Khrennikova, Dina, "Why the World Worries About Russia's Natural Gas Pipeline," www.Bloomberg.com, June 13, 2019. Updated on December 23, 2019. https://www.bloomberg.com/news/articles/2019-06-13/why-world-worries-about-russia-s-natural-gas-pipeline-quicktake
52   Harris, Tom, Dr. Lehr, Jay, "Science's Untold Scandal: The Lockstep March of Professional Societies to Promote the Climate Change Scare," www.PJMedia.com, May 24, 2019. https://pjmedia.com/news-and-politics/sciences-untold-scandal-the-lockstep-march-of-professional-societies-to-promote-the-climate-change-scare/

CHAPTER ONE: *THE GREEN NEW DEAL FUTURE - THE WORLD WITHOUT FOSSIL FUELS*

53   Khoury, Martin November 6, 2018, The Manure Crisis of 1894 https://www.beliefmedia.com.au/manure-crisis-1894
54   Liddell, Stephen The Great Horse Manure Crisis of 1894, September 11, 2017 https://stephenliddell.co.uk/2017/09/11/the-great-horse-manure-crisis-of-1894/
55   Green New Deal https://en.wikipedia.org/wiki/Green_New_Deal
56   Congressman Bernie Sanders' GND https://berniesanders.com/en/issues/green-new-deal/
57   Pros of the GND, February 15, 2019, https://www.dailyastorian.com/opinion/columns/pro-con-is-the-green-new-deal-worth-its-price/

article 81240c96-308d-11e9-9f15-436048b71817.html?utm_
medium=social&utm_source=email&utm_campaign=user-share

58 Cons of the GND, February 15, 2019, https://www.dailyastorian.
com/opinion/columns/pro-con-is-the-green-new-deal-worth-its-
price/article 81240c96-308d-11e9-9f15-436048b71817.html?utm_
medium=social&utm_source=email&utm_campaign=user-share

59 The GND is expensive, March 27, 2019, https://mises.org/
wire/study-estimates-green-new-deal-cost-93-trillion-%E2%80
%94-thats-conservative-estimate

60 One Third of World's Energy Could Be Solar by 2060, Predicts
Historically Conservative IEA, December 2, 2011, https://www.
renewableenergyworld.com/2011/12/02/one-third-of-worlds-energ
y-could-be-solar-by-2060-predicts-historically-conservative-iea/

61 What is the Paris Agreement? https://unfccc.int/process-and-meetings/
the-paris-agreement/what-is-the-paris-agreement

62 The Pros and Cons of the Paris Agreement https://ourglobalclimate.
com/pros-and-cons-of-paris-climate-agreement/

63 The Pros and Cons of the Paris Agreement https://ourglobalclimate.
com/pros-and-cons-of-paris-climate-agreement/

64 Clifford Krauss, Clifford, Oct. 29, 2017, Russia Uses Its Oil
Giant, Rosneft, as a Foreign Policy Tool https://www.nytimes.
com/2017/10/29/business/energy-environment/russia-venezula-oil-
rosneft.html

65 Krauss, Clifford, "Russia Uses Its Oil Giant, Rosneft, as a Foreign
Policy Tool," www.NYTimes.com, (New York Times), October
29, 2017. https://www.nytimes.com/2017/10/29/business/energy-
environment/russia-venezula-oil-rosneft.html

66 https://www.amazon.com/dp/1796049832/ Footnote 573 from Energy
Made Easy https://www.amazon.com/dp/1796049832/ Clifford
Krauss, Clifford, Oct. 29, 2017, Russia Uses Its Oil Giant, Rosneft, as
a Foreign Policy Tool https://www.nytimes.com/2017/10/29/business/
energy-environment/russia-venezula-oil-rosneft.html

67 The Economist Intelligence Unit, Russia Energy, "More than oil, gas
is Russia's main strategic asset," www.EIU.com, December 11, 2017.
Link gives amount proven, recoverable Russian energy reserves of oil
and natural gas to use as a leveraged weapon against the west. http://
www.eiu.com/industry/article/816211465/more-than-oil-gas-is-
russias-main-strategic-asset/2_3

68 Simple English Wikipedia, "Geopolitics," www.Wikipedia.com, Page accessed January 5, 2019. https://simple.wikipedia.org/wiki/Geopolitics

69 Blank, Stephen, "Russia has weaponized the energy sector in war against the West," www.TheHill.com, October 17, 2017. https://thehill.com/opinion/international/355742-russias-has-weaponized-the-energy-sector-in-war-against-the-west

70 Lauren Effron, Andrew Paparella, and Jeca Taudte, December 20, 2019, The scandals that brought down the Bakkers, once among US's most famous televangelists, https://abcnews.go.com/US/scandals-brought-bakkers-uss-famous-televangelists/story?id=60389342

71 Service, Robert, Can the world make the chemicals it needs without oil?, September. 19, 2019,https://www.sciencemag.org/news/2019/09/can-world-make-chemicals-it-needs-without-oil

72 Roger Pielke, Roger, October 27, 2019, The World Is Not Going To Halve Carbon Emissions By 2030, So Now What? https://www.forbes.com/sites/rogerpielke/2019/10/27/the-world-is-not-going-to-reduce-carbon-dioxide-emissions-by-50-by-2030-now-what/#407a74293794

73 https://www.cell.com/one-earth/fulltext/S2590-3322(19)30219-2? returnURL=https%3A%2F%2Flinkinghub.elsevier.com%2Fretrieve%2Fpii%2FS2590332219302192%3Fshowall%3Dtrue

74 Parts of a Wind turbine https://www.horizoncurriculum.com/supportmaterial/parts-of-a-wind-turbine/

75 Parts of a Solar System https://www.cleanenergyauthority.com/solar-energy-resources/components-of-a-residential-solar-electric-system

76 Reuters, June 27, 2019, U.S. dependence on China's rare earth: Trade war vulnerability, https://www.reuters.com/article/us-usa-trade-china-rareearth-explainer/u-s-dependence-on-chinas-rare-earth-trade-war-vulnerability-idUSKCN1TS3AQ

77 Conca, James, September 26, 2018 Blood Batteries - Cobalt And The Congo https://www.forbes.com/sites/jamesconca/2018/09/26/blood-batteries-cobalt-and-the-congo/#3daa5750cc6e

78 Minerals in Your Life http://mineralseducationcoalition.org/mining-minerals-information/minerals-in-your-life/

79 Bell, Terence, November 20, 2019, The World's Biggest Cobalt Refiners https://www.thebalance.com/the-biggest-cobalt-producers-2339726

80 The Top Lithium Producing Countries In The World https://www. worldatlas.com/articles/the-top-lithium-producing-countries-in-the-world.html

81 November 15, 2017, Industry giants fail to tackle child labour allegations in cobalt battery supply chains https://www. amnesty.org/en/latest/news/2017/11/industry-giants-fail-to-tackl e-child-labour-allegations-in-cobalt-battery-supply-chains/

82 cc Business News June 3, 2019, Explainer: U.S. dependence on China's rare earth - Trade war vulnerability. https://www.reuters.com/article/ us-usa-trade-china-rareearth-explainer/explainer-u-s-dependence-on-chinas-rare-earth-trade-war-vulnerability-idUSKCN1T42RP

83 West, Carl July 29, 2017 Carmakers' electric dreams depend on supplies of rare minerals https://www.theguardian.com/environment/2017/ jul/29/electric-cars-battery-manufacturing-cobalt-mining

84 November 15, 2017, Industry giants fail to tackle child labour allegations in cobalt battery supply chains https://www. amnesty.org/en/latest/news/2017/11/industry-giants-fail-to-tackl e-child-labour-allegations-in-cobalt-battery-supply-chains/

85 Figure 1-4 Wikimedia Commons, the free media repository, File:Child labor, Artisan Mining in Kailo Congo.jpg https://commons.wikimedia. org/wiki/File:Child labor, Artisan Mining in Kailo Congo.jpg

86 Armstrong, Paul, December 5, 2011, What are 'conflict diamonds?' https://www.cnn.com/2011/12/05/world/africa/conflict-diamonds-explainer/index.html

87 Spence, Katie, October 29, 2018 Tesla Motors' Dirty Little Secret Is a Major Problem https://www.fool.com/investing/general/2014/01/1 9/tesla-motors-dirty-little-secret-is-a-major-proble.aspx

88 Paul, Justin, EPA Document 908R11003, April 15, 2011 https:// nepis.epa.gov/Exe/ZyNET.exe/P100FHSY.txt?ZyActionD=Zy Document&Client=EPA&Index=2011%20Thru%202015&Docs =&Query=&Time=&EndTime=&SearchMethod=1&TocRestrict =n&Toc=&TocEntry=&QField=&QFieldYear=&QFieldMonth= &QFieldDay=&UseQField=&IntQFieldOp=0&ExtQFieldOp=0& XmlQuery=&File=D%3A%5CZYFILES%5CINDEX%20DATA% 5C11THRU15%5CTXT%5C00000006%5CP100FHSY.txt&User =ANONYMOUS&Password=anonymous&SortMethod=h%7C- &MaximumDocuments=1&FuzzyDegree=0&ImageQuality=r75 g8/r75g8/x150y150g16/i425&Display=hpfr&DefSeekPage=x&

SearchBack=ZyActionL&Back=ZyActionS&BackDesc=Results%20
page&MaximumPages=20&ZyEntry=1&slide

89 THIS IS WHAT WE DIE FOR: HUMAN RIGHTS ABUSES IN
THE DEMOCRATIC REPUBLIC OF THE CONGO POWER
THE GLOBAL TRADE IN COBALT January 15, 2016, https://
www.amnestyusa.org/reports/this-is-what-we-die-for-huma
n-rights-abuses-in-the-democratic-republic-of-the-congo-
power-the-global-trade-in-cobalt/

90 November 15, 2017, Industry giants fail to tackle child labour
allegations in cobalt battery supply chains https://www.
amnesty.org/en/latest/news/2017/11/industry-giants-fail-to-tackl
e-child-labour-allegations-in-cobalt-battery-supply-chains/

91 Naveena Sadasivam, Naveena, Apr 17, 2019, Going 100% renewable
power means a lot of dirty mining, https://grist.org/article/report-goin
g-100-renewable-power-means-a-lot-of-dirty-mining/

92 Institute for Energy Research, Obama Allows Wind Turbines to Legally
Kill Eagles January 9, 2017, https://www.instituteforenergyresearch.
org/renewable/wind/obama-allows-wind-turbines-legally-kill-eagles/

93 William Wilkes, Hayley Warren and Brian Parkin, August
15, 2018, Germany's Failed Climate Goals A Wake-Up Call
for Governments Everywhere https://www.bloomberg.com/
graphics/2018-germany-emissions/

94 Ellen Thalman, Benjamin Wehrmann, April1, 2019, What German
households pay for power, https://www.cleanenergywire.org/
factsheets/what-german-households-pay-power

95 AXELROD, Tal December 31, 2019, Merkel vows Germany will
do 'everything humanly possible' to fight climate change, https://
thehill.com/policy/energy-environment/476416-merkel-vows-germa
ny-will-do-everything-humanly-possible-to-fight

96 William Wilkes, Hayley Warren and Brian Parkin, August
15, 2018, Germany's Failed Climate Goals A Wake-Up Call
for Governments Everywhere https://www.bloomberg.com/
graphics/2018-germany-emissions/

97 Stevens, Pippa, November 13, 2019, Global energy demand means
the world will keep burning fossil fuels, International Energy
Agency warns, https://www.cnbc.com/2019/11/12/global-energ
y-demand-will-keep-world-burning-fossil-fuels-agency-says.html

98 Kiley, Sam, CNN, July 18, 2018, Vladimir Putin must be delighted
with his useful idiots in the West, https://www.cnn.com/2018/07/18/

opinions/vladimir-putin-and-his-useful-idiots-opinion-intl/index.
html

99   Stein, Ronald, September 24, 2018, America is following Germany's
     failed climate goals, https://www.cfact.org/2018/09/24/america-i
     s-following-germanys-failed-climate-goals/

100 Stein, Ronald, May 29, 2019, Australia's voters reject environmental
     fantasies, http://www.foxandhoundsdaily.com/2019/05/australia
     s-voters-reject-environmental-fantasies/

101 July 11, 2019, Pricing Vanity: Counting the Crushing Costs of
     Chaotically Intermittent Wind & Solar, https://stopthesethings.
     com/2019/07/11/pricing-vanity-counting-the-crushing-costs-of-ch
     aotically-intermittent-wind-solar/

102 Stein, Ronald, February 10, 2019, The Green New Deal would takes
     us back to medieval times, https://www.cfact.org/2019/02/10/the-gree
     n-new-deal-would-takes-us-back-to-medieval-times/#30a36ff1e4de

103 William Wilkes, Hayley Warren and Brian Parkin, August
     15, 2018, Germany's Failed Climate Goals A Wake-Up Call
     for   Governments   Everywhere   https://www.bloomberg.com/
     graphics/2018-germany-emissions/

104 Mundahl, Erin, March 12, 2019, US Still Subsidizing Renewable Energy
     to the Tune of Nearly $7 Billion, https://www.insidesources.com/
     us-still-subsidizing-renewable-energy-to-the-tune-of-nearly-7-billion/

105 Coal   Plants   by   Country   https://docs.google.com/spreadsheets/
     d/1I8GeKEfxPpwkQ_t0GQZx1GQm6MASclEtEtrQX3Y1nNc/
     edit#gid=0

106 Kieran Corcoran, Kieran, May 5, 2018, California's economy is now
     the 5th-biggest in the world, and has overtaken the United Kingdom,
     https://www.businessinsider.com/california-economy-ranks-5th-i
     n-the-world-beating-the-uk-2018-5

107 WWEA Press Release, February 25, 2019, Wind Power Capacity
     Worldwide Reaches 597 GW, 50,1 GW added in 2018, https://
     wwindea.org/blog/2019/02/25/wind-power-capacity-worldwide
     -reaches-600-gw-539-gw-added-in-2018/

108 PRESS RELEASE Newswire, October 1, 2019, Ronald Stein and
     Todd Royal announce the release of 'Energy Made Easy' https://
     markets.businessinsider.com/news/stocks/ronald-stein-and-todd-roya
     l-announce-the-release-of-energy-made-easy-1028565336

109 Shah, Anup, January 7, 2013, Poverty Facts and Stats, http://www.
     globalissues.org/article/26/poverty-facts-and-stats

110 UNICEF, Reduce Infant Mortality, https://static.unicef.org/mdg/childmortality.html

111 Worldometer, Current World Population, https://www.worldometers.info/world-population/

112 United Nations, Department of Economic and Social Affairs, June 21, 2017, World population projected to reach 9.8 billion in 2050, and 11.2 billion in 2100, https://www.un.org/development/desa/en/news/population/world-population-prospects-2017.html

113 UNICEF, Reduce Infant Mortality, https://static.unicef.org/mdg/childmortality.html

114 Voelcker, John, July 29, 2014, https://www.greencarreports.com/news/1093560_1-2-billion-vehicles-on-worlds-roads-now-2-billion-by-2035-report

115 Morris, Hugh, August 16, 2017, How many planes are there in the world right now? https://www.telegraph.co.uk/travel/travel-truths/how-many-planes-are-there-in-the-world/

116 Ranken Energy Corporation, Products made from petroleum, https://www.ranken-energy.com/index.php/products-made-from-petroleum/

117 EIA, Oil: crude and petroleum products explained, https://www.eia.gov/energyexplained/oil-and-petroleum-products/#tab1

118 White, Chris, August 27, 2018, Officials Issue Winter Weather Warnings For The Rockies — In Summer, https://dailycaller.com/2018/08/27/national-weather-service-wyoming-winter-snow/

119 Bastasch, Michael, August 6, 2019, National Park Removed Warning Glaciers 'Will All Be Gone' By 2020 After Years Of Heavy Snowfall, https://www.thegwpf.com/national-park-removed-warning-that-glaciers-will-all-be-gone-by-2020-after-years-of-heavy-snowfall/

120 Where's the beef ad, May 12, 2006, https://www.youtube.com/watch?v=Ug75diEyiA0

121 Climate Action Network, **USCAN is a vital network for 175+ organizations active on climate change,** https://www.usclimatenetwork.org/member-organizations

122 The Spirit of Democratic Capitalism, https://en.wikipedia.org/wiki/The_Spirit_of_Democratic_Capitalism

123 Smith, Adam, *An Inquiry Into the Nature and Causes of the Wealth of Nations*, (W. Strathan and T. Cadell, London, England), Entire book. Original Publication Date: 1776. https://www.amazon.com/Inquiry-Nature-Causes-Wealth-Nations-ebook/dp/B00847CE6O/ref=sr_1_2?keywords=adam+smith&qid=1578350978&sr=8-2

124 Wikipedia.com, "Definition of Mercantilism," www.Wikipedia.com, Page accessed on January 6, 2020. https://en.wikipedia.org/wiki/Mercantilism

125 Hazlitt, Henry, *The Conquest of Poverty*, (Arlington House, Rochelle, N.Y.), 1973 and online the date is February 6, 2012 (updated version), Pages 13-18. https://www.amazon.com/Conquest-Poverty-Henry-Hazlitt-ebook/dp/B0076DGJ7M/ref=sr_1_1?keywords=the+conquest+of+poverty&qid=1578351861&sr=8-1

126 Please see Paul Johnson, "Has Capitalism a Future?" in *Will Capitalism Survive?* ed. Ernest W. Lefever (Washington, D.C.: Ethics and Public Policy Center, 1979), page 5.

127 Webster, Ian, CPI Inflation Calculator, "Inflation Calculator," www.OficialData.org, January 15, 2020. Data originates from U.S. Bureau of Labor Statistics consumer price index. https://www.officialdata.org/us/inflation/1800?amount=1000

128 Ibid. Stein and Royal. 2020. Please see introduction to the book for source verification.

129 Stein, Ronald, "Third World Countries Remain The Losers of Climate Change Activism," www.EurasiaReview.com, November 21, 2019. https://www.eurasiareview.com/21112019-third-world-countries-remain-the-losers-of-climate-change-activism-oped/

130 Stein. Ibid. 2019.

131 Novak. Ibid. Page 17. Entire paragraph. 1990.

132 Roberts, Andrew, *Churchill Walking With Destiny*, (Penguin Books, New York, NY), Page 39, November 6, 2018. https://www.amazon.com/s?k=churchill+walking+with+destiny+by+andrew+roberts&crid=AJ5LQDKY3GRG&sprefix=%2Caps%2C270&ref=nb_sb_ss_i_1_0

133 Hankyoreh Hani.co.kr posting, "From darkness to light: North Koreans experience abundance of electricity for first time," www.

english.hani.co.kr, January 14, 2019. http://english.hani.co.kr/arti/english_edition/e_northkorea/878329.html

134 Novak, Michael, *The Spirit of Democratic Capitalism*, (Madison Books, Lanham, MA), Introduction, pages 14. December 29, 1990. Entire paragraph is sourced from the entire book. https://www.amazon.com/Spirit-Democratic-Capitalism-Michael-Novak/dp/0819178233/ref=sr_1_1?ie=UTF8&qid=1545324496&sr=8-1&keywords=the+spirit+of+democratic+capitalism

135 Novak. Ibid. Page 17. 1990.

136 Hayek, F.A., "The Standard of Life of the Workers in England," an excerpt from *Capitalism and the Historians*, (Chicago Press, University of Chicago, Chicago, IL), Pages 152-54, 1954.

137 Tracy Ellis, John, *American Catholicism*, (University of Chicago Press, Chicago, IL), Page 106, August 1969. https://www.amazon.com/American-Catholicism-History-Civilization/dp/0226205541/ref=sr_1_2?keywords=american+catholicism+john+tracy+ellis&qid=1578428690&sr=8-2

138 Jenkins, Phillip, *The Next Christendom: The Coming of Global Christianity*, (Oxford University Press, Oxford, England), Entire book. September 13, 2011. https://www.amazon.com/Next-Christendom-Coming-Global-Christianity/dp/0199767467/ref=sr_1_1?crid=2297RYXFKXWDV&keywords=the+next+christendom&qid=1578429117&sprefix=the+next+ch%2Caps%2C199&sr=8-1

139 Dinan, Stephen, "Losing our Religion: America becoming 'pagan' as Christianity cedes to culture," www.WashingtonTimes.com, December 30, 2019. https://www.washingtontimes.com/news/2019/dec/30/faith-in-us-withers-as-apathy-trumps-religion/

140 Clark, Heather, "Planned Parenthood Murdered Record Number of Unborn Babies in 2018-2019 While Receiving $616 Million in Taxpayer Funds," www.ChristianNews.net, January 7, 2019.

141 Royal, Todd, "The Unintended Geopolitical Consequences of Abortion," www.CaPoliticalReview.com, January 5, 2017. http://www.capoliticalreview.com/top-stories/the-unintended-geopolitical-consequences-of-abortion/

142 Nisbet, Robert, *History of the Idea of Progress*, 2nd edition, (Routledge Taylor & Francis Group, Philadelphia, PA), Entire Book, January 31, 1994. https://www.amazon.com/History-Idea-Progress-Robert-Nisbet/dp/1560007133/ref=sr_1_1?crid=2UJMUBUCIM4KR

&keywords=history+of+the+idea+of+progress&qid=157842982
5&sprefix=history+of+the+idea+of+pr%2Caps%2C196&sr=8-1

143 Kagan, Donald, *Pericles Of Athens And The Birth Of Democracy*,
(Free Press, New York, NY), October 1, 1998. https://www.amazon.
com/Pericles-Athens-Birth-Democracy-Donald/dp/0684863952/
ref=sr_1_1?keywords=pericles&qid=1578528783&sr=8-1

144 Gibbon, Edward, *The History of the Decline and Fall of the
Roman Empire, All Six Volumes in One Book*, (Amazon Services
LLC, Seattle, WA), December 18, 2013. https://www.amazon.
com/History-Decline-Roman-Empire-Volumes-ebook/dp/B00HIM
09MM/ref=sr_1_3?crid=2E6ZRH28ARLJ3&keywords=the+r
ise+and+fall+of+the+roman+empire&qid=1578349272&sp
refix=the+rise+and+fall+of+the+roman+em%2Caps%2C191&sr=8-3

145 Novak. Ibid. Page 14. 1990.

146 Novak. Ibid. Page 14. 1990.

147 United Nations, United Nations Universal Declaration of Individual
Rights, www.UN.org, First proclaimed in Paris, France, December
10, 1948. https://www.un.org/en/universal-declaration-human-rights/

148 Epstein, Alex, *The Moral Case for Fossil Fuels*, (Portfolio–Penguin Books,
New York, NY), Entire book for the source. November 13, 2014. https://
www.amazon.com/Moral-Case-Fossil-Fuels/dp/1591847443/ref=sr_
1_1?ie=UTF8&qid=1544553425&sr=8-1&keywords=alex+epstein

149 Smil, Vaclav, *Energy and Civilization A History*, (The MIT Press,
Cambridge, MA), Entire book for the source. May 12, 2017.
https://www.amazon.com/Energy-Civilization-History-MIT-
Press/dp/0262035774/ref=redir_mobile_desktop?
encoding=UTF8&aaxitk=5fms2Q956nPjIpYAcugwmg&hsa_cr
id=2643231280501&ref_=sb_s_sparkle_slot

150 Chrenkoff, Arthur, "We don't just have a bushfire crisis. We have an
arson crisis, too," www.Spectator.com.au, January 4, 2020. https://
www.spectator.com.au/2020/01/we-dont-just-have-a-bushfire-cris
is-we-have-an-arson-crisis-too/

151 Betigeri, Aarti, "How Australia's Indigenous Experts Could Help
Deal With Devastating Wildfires," www.Time.com, January 14,
2020. Source is for final two sentences of the paragraph. https://time.
com/5764521/australia-bushfires-indigenous-fire-practices/

152 Driessen, Paul, "Reform USAID Energy Aid Polices ASAP!" www.
Townhall.com, January 4, 2020. Article co-authored David Wojick,
which was included as an editor's note. https://townhall.com/

columnists/pauldriessen/2020/01/04/reform-usaid-energy-aid-policie
s-asap-n2558947

153 Thomas, Tony, "Maths is Hard for the Green-Minded," www.
Quadrant.org.au, October 15, 2019. https://quadrant.org.au/opinion/
doomed-planet/2019/10/maths-is-hard-for-the-green-minded/

154 Roberts. Ibid. Page 39-44. 2018.

155 Driessen, Paul, Wojick, David, "Climate alarmists banks go carbon-
colonialist," www.Cfact.org, December 30, 2019. https://www.cfact.
org/2019/12/30/climate-alarmist-banks-go-carbon-colonialist/

156 Driessen, Wojick. Ibid. 2020.

157 Launch of the Japan-Africa Energy Initiative, Remarks by Akinwumi
Adesina, President of the African Development Bank, Letter of Intent
signing ceremony at the Africa Union Summit, Addis Ababa, July 3, 2017.
www.AFDB.org. https://www.afdb.org/fileadmin/uploads/afdb/Docu
ments/Generic-Documents/Launch of the Japan - Africa Energy
Initiative - Remarks by Akinwumi Adesina President of the
African Development Bank - Letter of Intent signing ceremony
at the Africa Union Summit Addis Abab 3 July 2017.pdf

158 USAID, Climate Links, "Enhancing Capacity for Low Emission
Development Strategies," www.climatelinks.org, May 2017. Page
accessed on January 8, 2020. https://www.climatelinks.org/resources/
enhancing-capacity-low-emission-development-strategies-ec-leds

159 DoSomething.org, "11 Facts About Global Poverty," www.
DoSomething.org, Page accessed January 15, 2020. https://www.
dosomething.org/us/facts/11-facts-about-global-poverty

160 Driessen, Paul, "No Plan B for Planet A," www.Cfact.org,
November 25, 2019. https://wattsupwiththat.com/2019/11/25/
no-plan-b-for-planet-a/

161 Ridley, Matt, "Wind turbines are neither clean nor green and they
provide zero global energy," www.Spectator.Co.Uk, May 13, 2017.
https://www.spectator.co.uk/2017/05/wind-turbines-are-neithe
r-clean-nor-green-and-they-provide-zero-global-energy/

162 Jones, Barbara for the Mail on Sunday, "Child miners aged four living
a hell on Earth so YOU can drive an electric car: Awful human cost
in squalid Congo cobalt mine that Michael Gove didn't consider in
his 'clean' energy crusade," www.DailyMail.Co.Uk, August 5, 2017.
https://www.dailymail.co.uk/news/article-4764208/Child-miners-age
d-four-living-hell-Earth.html

163 Frankel, Todd C., "THE COBALT PIPELINE: Tracing the path from deadly hand-dug mines in Congo to consumers' phones and laptops," www.WashingtonPost.com, September 30, 2016. https://www.washingtonpost.com/graphics/business/batteries/congo-cobalt-mining-for-lithium-ion-battery/

164 UNICEF, United Nations organization, Millennium Development Goals, "Reduce child mortality," www.static.Unicef.org, Page accessed January 15, 2020. https://static.unicef.org/mdg/childmortality.html

165 Driessen, Wojick. Ibid. 2020.

166 Driessen, Wojick. Ibid. 2020.

167 Gronholt-Pedersen, Jacob, "Denmark sources record 47% of power from wind in 2019," www.Reuters.com, January 2, 2020. https://www.reuters.com/article/us-climate-change-denmark-windpower/denmark-sources-record-47-of-power-from-wind-in-2019-idUSKBN1Z10KE

168 BBC News, "Nobel Prize-winning scientist Frances Arnold retracts paper," www.BBC.com, January 3, 2020. https://www.bbc.com/news/world-us-canada-50989423

169 Kotkin, Joel, Toplansky, Marshall, "Kotkin: California becoming more feudal, with ultra-rich lording over declining middle class," www.OCRegister.com, October 14, 2018. https://www.ocregister.com/2018/10/14/kotkin-california-becoming-more-feudal-with-ultra-rich-lording-over-declining-middle-class/

170 Soldatkin, Vladimir, "UPDATE 1-Russian oil, condensate output surges to record-high in 2019," www.Reuters.com, January 1, 2020. https://www.reuters.com/article/russia-energy-production/update-1-russian-oil-condensate-output-surges-to-record-high-in-2019-idUSL8N2970BX

171 Khrennikova, Dina, Rudnitsky, Jake, "Russia's Oil Output Hits Post –Soviet Record Despite OPEC+ Deal," www.BnnBloomberg.ca, January 2, 2020. https://www.bnnbloomberg.ca/russia-s-oil-output-hits-post-soviet-record-despite-opec-deal-1.1368250

172 Koutantou, Angeliki, "Greece, Israel, Cyprus sign EastMed gas pipeline deal," www.UK.Reuters.com, January 2, 2020. https://uk.reuters.com/article/uk-greece-cyprus-israel-pipeline/greece-israel-cyprus-sign-eastmed-gas-pipeline-deal-idUKKBN1Z116L

173 Royal, Todd, "U.S. Natural Gas is the new, Global, Soft Power Weapon," www.NationalInterest.org, August 27, 2019. https://nationalinterest.org/feature/us-natural-gas-new-global-soft-power-weapon-76471

174 Tanas, Olga, "Russia Readies New Gas Link to Europe in Defiance of U.S.," www.Bloomberg.com, January 7, 2020. https://www.bloomberg.com/news/articles/2020-01-08/russia-opens-natural-gas-link-to-turkey-amid-u-s-opposition

175 Gardner, Timothy, "U.S. greenhouse gas emissions dip, Trump policies put future cuts in doubt: study," www.Reuters.com, January 7, 2020. https://www.reuters.com/article/us-usa-emissions/u-s-greenhouse-gas-emissions-dip-trump-policies-put-future-cuts-in-doubt-study-idUSKBN1Z616Y

176 Desjardins, Jeff, Visual Capitalist, "All of these things can be made with one barrel of oil," www.BusinessInsider.com, September 28, 2016. https://www.businessinsider.com/things-that-can-be-made-with-one-barrel-of-oil-2016-9

177 Slav, Irina, "Why Big Tech Is Backing Big Oil," www.OilPrice.com, January 6, 2020. https://oilprice.com/Energy/Energy-General/Why-Big-Tech-Is-Backing-Big-Oil.html

178 Lieberman, Joe, "The Democrats and Iran: Why can't the party's candidates simply admit Qasem Soleimani's death makes America safer?" www.WSJ.com, January 5, 2020. https://www.wsj.com/articles/the-democrats-and-iran-11578262553?fbclid=IwAR3UcVQPVbq-hK-ZbD1PR-UQgpKAEaW3mKq4rbMeDvioJ7otXT0ucwqJKN8

179 Bailey, Ronald, *The End of Doom: Environmental Renewal in the Twenty-First Century*, (Thomas Dunn Books, Stuttgart, Germany) Entire Book. July 21, 2015. https://www.amazon.com/End-Doom-Environmental-Renewal-Twenty-first/dp/1250057671/ref=sr_1_1?keywords=the+end+of+doom&qid=1578433621&sr=8-1

CHAPTER THREE: *POST 1900'S ERA - AFTER OIL BEGAN TO SUPPORT AUTOMOBILES AND AIRPLANES*

180 Thousands of products from petroleum, https://en.wikipedia.org/wiki/Petroleum_product

181 Thousands of products from petroleum, https://en.wikipedia.org/wiki/Petroleum_product

182 Morris, Hugh, August 16, 2017, The Telegraph, https://www.telegraph.co.uk/travel/travel-truths/how-many-planes-are-there-in-the-world/

183 COUNTRY COMPARISON: **AIRPORTS,** Central Intelligence Agency, https://www.cia.gov/library/publications/resources/the-world-factbook/fields/379rank.html

184 Airports Council International, April 9, 2018, https://aci.aero/news/2018/04/09/aci-world-releases-preliminary-2017-world-airport-traffic-rankings-passenger-traffic-indian-and-chinese-airports-major-contributors-to-growth-air-cargo-volumes-surge-at-major-hubs-as-trade-wars-thre/

185 Traveller, Plane facts and truths, https://www.traveller.com.au/plane-facts-and-truths-how-many-aircraft-are-there-in-the-world-right-now-and-how-many-have-disappeared-gxy2xs

186 E. Mazareanu, Dec 11, 2019, Global air traffic - number of flights 2004-2020, https://www.statista.com/statistics/564769/airline-industry-number-of-flights/

187 World Jet Fuel Consumption per Year, https://www.indexmundi.com/energy/?product=jet-fuel

188 Travel Week News, Exactly how many planes are there in the world today?, February 17, 2017, http://www.travelweek.ca/news/exactly-many-planes-world-today/

189 Morris, Hugh, August 16, 2017, The Telegraph, https://www.telegraph.co.uk/travel/travel-truths/how-many-planes-are-there-in-the-world/

190 Cruise Market Watch, https://cruisemarketwatch.com/capacity/

191 Erin De Santiago, https://cruises.lovetoknow.com/wiki/How_Much_Fuel_Does_a_Cruise_Ship_Use

192 Oilprice.com, https://oilprice.com/Energy/Energy-General/Diesel-Demand-Is-Set-To-Soar.html

193 The Geography of Transport Systems, Fuel Consumption by Containership Size and Speed, https://transportgeography.org/?page_id=5955

194 Michael, U.S. Special Delivery Freight Solutions, February 23, 2017, https://www.usspecial.com/how-many-trucking-companies-in-the-usa/

195 Hsu, Tiffany, Trucks.com, October 17, 2016, https://www.trucks.com/2016/10/17/truckers-fuel-efficiency-alternative-fuels/

196 Mitchell, Russ, Car buyers shun electric vehicles, January 17, 2020, https://www.latimes.com/business/story/2020-01-17/ev-sales-fizzle?utm_source=newsletter&utm_medium=email&utm_campaign=newsletter_axiosgenerate&stream=top

---

197 EVAdoption, https://evadoption.com/ev-market-share/ev-market-share-california/ This plot is a graphical representation of the EV sales data set for the noted States shown as a graph showing the relationship between the State and Sales

198 Green Car Reports, https://www.greencarreports.com/news/1093560_1-2-billion-vehicles-on-worlds-roads-now-2-billion-by-2035-report

199 EIA, https://www.eia.gov/energyexplained/oil-and-petroleum-products/#tab1

200 Thousands of products from petroleum, https://en.wikipedia.org/wiki/Petroleum_product

201 EIA, Today in Energy, SEPTEMBER 14, 2017, https://www.eia.gov/todayinenergy/detail.php?id=32912

202 Energy.gov, 3 Reasons Why Nuclear is Clean and Sustainable, September 25, 2018, https://www.energy.gov/ne/articles/3-reasons-why-nuclear-clean-and-sustainable

203 Coal Plants by Country https://docs.google.com/spreadsheets/d/1I8GeKEfxPpwkQ_t0GQZx1GQm6MASclEtEtrQX3Y1nNc/edit#gid=0

204 Unicef, https://static.unicef.org/mdg/childmortality.html

205 Wikipedia, Petroleum refining in the United States, https://en.wikipedia.org/wiki/Petroleum_refining_in_the_United_States

206 Wikipedia, List of Refineries, https://en.wikipedia.org/wiki/List_of_oil_refineries

207 Testimony before the Subcommittee on Environment and Oversight, February 24, 2017, https://docs.house.gov/meetings/SY/SY18/20170228/105632/HHRG-115-SY18-Wstate-DayaratnaK-20170228.pdf

208 Loris, Nicolas, The Heritage Foundation, https://www.heritage.org/energy-economics/commentary/green-new-deal-would-barely-change-earths-temperature-here-are-the

209 HASEMYER, DAVID January 17, 2020, https://insideclimatenews.org/news/04042018/climate-change-fossil-fuel-company-lawsuits-timeline-exxon-children-california-cities-attorney-general

210 EIA, Today in Energy, SEPTEMBER 24, 2019, https://www.eia.gov/todayinenergy/detail.php?id=41433

211 Products from Petroleum, https://whgbetc.com/petro-products.pdf

212 Unicef, https://static.unicef.org/mdg/childmortality.html

213 Unicef, https://static.unicef.org/mdg/childmortality.html

214 Real time population source, https://www.worldometers.info/world-population/

215 Held, Amy, June 22, 2017, NPR, https://www.npr.org/sections/thetwo-way/2017/06/22/533935054/u-n-says-world-s-population-will-reach-9-8-billion-by-2050

216 Unicef, https://static.unicef.org/mdg/childmortality.html

217 Shah, Anup, January 7, 2013, Global Issue Poverty Facts and Stats, http://www.globalissues.org/article/26/poverty-facts-and-stats

CHAPTER FOUR: *UNDERDEVELOPED COUNTRIES FUTURE - CONTINUATION OF DEPRIVATION OF PRODUCTS FROM CRUDE OIL*

218 Ranken Energy Corporation, "Products made from petroleum: With Over 6000 products and counting, petroleum continues to be a crucial requirement for all consumers," www.Ranken-Energy.com, 2017. Page accessed January 14, 2020. https://www.ranken-energy.com/index.php/products-made-from-petroleum/

219 Smil, Vaclav, "To Get Wind Power You Need Oil," www.Spectrum.IEEE.org, February 29, 2016. https://spectrum.ieee.org/energy/renewables/to-get-wind-power-you-need-oil

220 Chen, C. Julian, "Physics of Solar Energy," Columbia University, Department of Mathematics, www.Columbia.edu, August 2010. Source is used since the study shows all facets of solar panels and technology that have their origins in a 42-gallon barrel of crude oil. http://www.columbia.edu/~jcc2161/documents/Solar_Energy.pdf

221 U.S. Energy Information Administration, Independent Statistics & Analysis, "New electric generating capacity in 2020 will come primarily from wind and solar," www.EIA.gov, January 14, 2020. https://www.eia.gov/todayinenergy/detail.php?id=42495

222 Kool, Tom, "Oversupply Fears Are Front And Center in Oil Markets," www.OilPrice.com, January 14, 2020. https://oilprice.com/Energy/Energy-General/Oversupply-Fears-Are-Front-And-Center-In-Oil-Markets.html

223 Shellenberger, Michael, "Why Renewables Can't Save the Planet," www.Quillette.com from Forbes Magazine, February 27, 2019. https://quillette.com/2019/02/27/why-renewables-cant-save-the-planet/

224 Shellenberger, Michael, "Unreliable Nature Of Solar And Wind Makes Electricity More Expensive, New Study Finds,"

www.Forbes.com, April 22, 2019. https://www.forbes.com/
sites/michaelshellenberger/2019/04/22/unreliable-nature-of-
solar-and-wind-makes-electricity-much-more-expensive
-major-new-study-finds/#5ee85a764f59

225 Mills, Mark P., "If You Want 'Renewable Energy,' Get Ready to
Dig: Building one wind turbine requires 900 tons of steel, 2,500
tons of concrete and 45 tons of plastic," www.WSJ.com, August 5,
2019. https://www.wsj.com/articles/if-you-want-renewable-energy-ge
t-ready-to-dig-11565045328

226 Schiffman, Richard, "COLUMN-Is nuclear power the answer on
climate change?" www.Reuters.com, January 10, 2014. https://www.
reuters.com/article/schiffman-nuclear-idUSL2N0KK2HX20140110

227 DiSavino, Scott, "U.S. coal-fired power plants closing fast despite
Trump's pledge of support for industry," www.Reuters.com, January
13, 2020. https://www.reuters.com/article/us-usa-coal-decline-
graphic/u-s-coal-fired-power-plants-closing-fast-despite-trumps
-pledge-of-support-for-industry-idUSKBN1ZC15A

228 U.S. Energy Information Administration, Independent Statistics &
Analysis, Today in Energy, "EIA's latest International Energy Outlook
highlights analysis of China, India and Africa," www.EIA.gov, July
24, 2018. https://www.eia.gov/todayinenergy/detail.php?id=36732

229 Mills, Mark P., "The "New Energy Economy": An Exercise
in Magical Thinking," www.Manhattan-Institute.org,
March 26, 2019. https://www.manhattan-institute.org/
green-energy-revolution-near-impossible

230 McCarthy, Niall, "Report: India Lifted 271 Million
People Out Of Poverty In A Decade [Infographic]," www.
Forbes.com, July 12, 2019. https://www.forbes.com/sites/
niallmccarthy/2019/07/12/report-india-lifted-271-million-people
-out-of-poverty-in-a-decade-infographic/#2b1803852284

231 Romel, Valentina, Reed, John, "The Asian century is set to
begin," www.FT.com, March 25, 2019. https://www.ft.com/
content/520cb6f6-2958-11e9-a5ab-ff8ef2b976c7

232 Royal, Todd, "Energy Storage Isn't Ready For Wide Deployment,"
www.iagenda21.com via www.NewsGeography.com, November
20, 2018. http://iagenda21.com/energy-storage-isnt-ready-fo
r-wide-deployment/

233 Ebell, Myron, Milloy, Steven J., "Wrong Again: 50 Years of Failed
Eco-pocalyptic Predictions," www.CEI.org (Competitive Enterprise

Institute), September 18, 2019. https://cei.org/blog/wrong-again-5
0-years-failed-eco-pocalyptic-predictions

234 Elliott, Stuart, "UK oil, gas industry 'social license to operate'
under threat: OCA chairman," www.SPGlobal.com, January 16,
2020. https://www.spglobal.com/platts/en/market-insights/latest-
news/coal/011620-uk-oil-gas-industry-social-license-to-operate-un
der-threat-oga-chairman

235 Royal, Todd, "China and India Will Watch the West Destroy Itself,"
www.ModernDiplomacy.eu, June 5, 2019. https://moderndiplomacy.
eu/2019/06/05/china-and-india-will-watch-the-west-destroy-itself/

236 Maizland, Lindsay, "China's Repression of Uighurs in Xinjiang: More
than a million Muslims have been arbitrarily detained in China's
Xinjiang Province. The reeducation camps are just one part of the
government's crackdown on Uighurs," www.CFR.org (Council on
Foreign Relations), November 25, 2019. https://www.cfr.org/backgr
ounder/chinas-repression-uighurs-xinjiang

237 Maizland. Ibid. 2019

238 Saltskog, Mollie, Clarke, Colin P., "China's Rights Abuses
in Xinjiang Could Provoke a Global Terrorist Backlash,"
www.ForeignAffairs.com, January 16, 2020. https://www.
foreignaffairs.com/articles/china/2020-01-16/chinas-rights-
abuses-xinjiang-could-provoke-global-terrorist-backlash?utm
medium=newsletters&utm source=fatoday&utm
content=20200116&utm campaign=FA%20Today%20
011620%20Addressing%20China%27s%20Abuses%20in%20
Xinjiang%2C%20How%20to%20Reform%20Capitalism%2C%20
Female%20Politicians%20Under%20Attack&utm term=FA%20
Today%20-%20112017

239 U.S. Energy Information Administration, Independent Statistics &
Analysis, Today in Energy, "EIA's latest International Energy Outlook
highlights analysis of China, India, and Africa," www.EIA.gov, July
24, 2018. https://www.eia.gov/todayinenergy/detail.php?id=36732

240 Inskeep, Steve, Westerman, Ashley, "Why Is China Placing A
Global Bet On Coal?" www.NPR.org (U.S. National Public Radio),
April 29, 2019. https://www.npr.org/2019/04/29/716347646/
why-is-china-placing-a-global-bet-on-coal

241 Ebrahim, Zoleen, "Why is Pakistan still pursuing coal?" www.
ChinaDialogue.net, July 18, 2019. https://www.chinadialogue.net/
article/show/single/en/11388-Why-is-Pakistan-still-pursuing-coal-

242 Davidson, John Daniel, "A Drug Cartel Just Defeated The Mexican Military In Battle," www.TheFederalist.com, October 21, 2020. https://thefederalist.com/2019/10/21/a-drug-cartel-just-defeated-the-mexican-military-in-battle/

243 Cunningham, Nick, "The Next Oil Boom Is Happening Here," www.OilPrice.com, December 29, 2019. https://oilprice.com/Energy/Crude-Oil/The-Next-Oil-Boom-Is-Happening-Here.html

244 Mead, Walter Russell, "How American Fracking Changes the World," www.WSJ.com, November 26, 2018. https://www.wsj.com/articles/how-american-fracking-changes-the-world-1543276935

245 Morningstar, Cory, "The Manufacturing of Greta Thunberg – for Consent The Political Economy of the Non-Profit Industrial Complex," www.TheArtofAnnihilation.com, January 17, 2019. http://www.theartofannihilation.com/the-manufacturing-of-greta-thunberg-for-consent-the-political-economy-of-the-non-profit-industrial-complex/

246 Vaughn, Elizabeth, "Busted! Facebook Bug Reveals Greta Thunberg's Posts Are Written by Her Father," www.RedState.com, January 13, 2020. https://www.redstate.com/elizabeth-vaughn/2020/01/13/facebook-bug-reveals-greta-thunbergs-posts-are-written-by-her-father/

247 Harris, Emma, "How to Stop Freaking Out and Tackle Climate Change," www.NYTimes.com, January 10, 2020. https://www.nytimes.com/2020/01/10/opinion/sunday/how-to-help-climate-change.html

248 Mufson, Steve, "Bloomberg unveils plan to make new buildings 'zero-carbon' by 2025, www.Newsbreak.com, via The Washington Post, January 15, 2020. The above two paragraphs are for this source. https://www.newsbreak.com/news/0NqgiatK/bloomberg-unveils-plan-to-make-new-buildings-zero-carbon-by-2025

249 Banerjee, Abhijit V., Duflo, Esther, "How Poverty Ends," www.ForeignAffairs.com (Council on Foreign Relations), January/February 2020. https://www.foreignaffairs.com/articles/2019-12-03/how-poverty-ends?utm_medium=newsletters&utm_source=fatoday&utm_content=20200122&utm_campaign=FA%20Today%20012220%20The%20Cost%20of%20an%20Incoherent%20Foreign%20Policy%2C%20How%20Poverty%20Ends%2C%20The%20WHO%20and%20Epidemics&utm_term=FA%20Today%20-%20112017

250 Myers Jaffe, Amy, "Striking Oil Ain't What It Used to Be: Poor Countries Find Fossil Fuels Just as the Rich World Swears Them Off," www.ForeignAffairs.com, January 20, 2020. https://www.foreignaffairs.com/articles/africa/2020-01-20/striking-oil-aint-what-it-used-be

251 Rogers, Norman, "Another Expensive Solar Scheme Bites the Dust," www.AmericanThinker.com, January 8, 2020. https://www.americanthinker.com/articles/2020/01/another_expensive_solar_scheme_bites_the_dust.html

252 Rogers. Ibid. 2020.

253 McKinsey & Co., lead partner on the report, Kirsten Best-Werbunat, *Energiewende-Index*, www.McKinsey.de, August 2019. https://www.mckinsey.de/branchen/chemie-energie-rohstoffe/energiewende-index

254 Shellenberger, Michael, "Renewables Threaten Germany Economy & Energy Supply, McKinsey Warns In New Report," www.Forbes.com, September 5, 2019. https://www.forbes.com/sites/michaelshellenberger/2019/09/05/renewables-threaten-german-economy-energy-supply-mckinsey-warns-in-new-report/#3546cecc8e48

255 von Dohmen, Frank, Jung, Alexander, Schultz, Stefan, Traufetter, Gerald, "German Failure on the Road to a Renewable Future," Climate Stasis section, www.Spiegel.de, May 13, 2019. https://www.spiegel.de/international/germany/german-failure-on-the-road-to-a-renewable-future-a-1266586.html

256 Lomberg, Bjorn, "How Climate Policies Hurt the Poor," www.ProjectsSyndicate.com, September 26, 2019. https://www.project-syndicate.org/commentary/governments-must-reduce-poverty-not-emissions-by-bjorn-lomborg-2019-09?fbclid=IwAR1yAFeW-qGp76jo0fWmOX9r_Nyqd_6DNJabyh6PAgwlJBL-NiF5Bmilatc

257 Appunn, Kerstine, Wettengel, Julian, "Germany's greenhouse gas emissions and climate targets," www.CleanEnergyWire.org, January 7, 2020. https://www.cleanenergywire.org/factsheets/germanys-greenhouse-gas-emissions-and-climate-targets

258 Berlin, J.C., *The Economist* explains, "Why Germany's army is in a bad state," www.Economist.com, August 9, 2018. https://www.economist.com/the-economist-explains/2018/08/09/why-germanys-army-is-in-a-bad-state

259 European Commission, *The Commission presents strategy for a climate neutral Europe by 2050 – Questions and answers*, (European

Union, Brussels, Belgium), November 2018. https://ec.europa.eu/commission/presscorner/detail/en/MEMO_18_6545

260 Knopf, Brigitte, Yen-Heng Chen, Henry, De Cian, Enrica, Forster, Hannah, Kanudia, Amit, Karkatsouli, Ioanna, Keppo, Ilkka, Koljion, Tina, P. Van Vuuren, Detlef, *Beyond 2020 – Strategies And Costs For Transforming The European Energy System*, www.WorldScientific.com, Climate Change Economics, Vol. 04, No. supp01, 1340001, 2013. https://www.worldscientific.com/doi/abs/10.1142/S2010007813400010

261 Lomberg. Ibid. 2020.

262 Ballingail, Dr. Pambudi, Daniel, Dr. Corong, Erwin, Dr. Stroombergen, Adolf, Clough, Peter, New Zealand Institute of Economic Research (NZIER), *Economic impact analysis of 2050 emissions targets*, www.MFE.Govt.NZ, June 18, 2018. This source is for the entire paragraph. https://www.mfe.govt.nz/sites/default/files/media/Climate%20Change/NZIER%20report%20-%20Economic%20impact%20analysis%20of%202050%20emissions%20targets%20-%20FINAL.pdf

263 Veysey, Jason, Octaviano, Claudia, Calvin, Katherine, Hereras Martinize, Sara, Kitous, Alban, McFarland, James, van der Zwaan, Bob, *Pathways to Mexico's climate change mitigation targets: A multi-model analysis*, Energy Economics, Volume 56, Pages 587-599, May 2016. https://www.sciencedirect.com/science/article/pii/S0140988315001346?via%3Dihub

264 United Nations Climate Change, 2020 United Nations Framework Convention on Climate Change, *The Paris Agreement*, www.UNFCC.int, 2020. https://unfccc.int/process-and-meetings/the-paris-agreement/the-paris-agreement

265 Lomberg. Ibid. 2019.

266 Campagnolo, Lorenza, Davide, Marinella, "Can the Paris deal boost SDGs achievement? An assessment of climate mitigation co-benefits or side-effects on poverty and inequality," www.ScienceDirect.com, *World Development*, Volume 122, Pages 96-109, October 2019. https://www.sciencedirect.com/science/article/abs/pii/S0305750X19301299

267 Riahi, Keywan, P. van Vuuren, Detlef, Kriegler, Elmar, Edmonds, Jae, C. O'Neill, Brian, Fujimori, Shinichiro, Bauer, Nico, Calvin, Katherine, Dellink, Rob, Fricko, Oliver, Lutz, Wolfgang, Popp, Alexander, Crespo Cuaresma, Jesus, KC, Samir, Leimbach, Marian, Jiang, Leiwen, Kram, Tom, Rao, Shilpa, *The Shared Socioeconomic Pathways and*

their energy, land use, and greenhouse gas emissions implications: An overview, www.ScienceDirect.com, *Global Environmental Change*, Volume 42, Pages 153-168, January 2017. https://www.sciencedirect.com/science/article/pii/S0959378016300681

268 Lomberg. Ibid. 2019.

269 D. Rao, Narasimha, Sauer, Petra, Gidden, Matthew, Riahi, Keywan, *Income inequality projections for the Shared Socioeconomic Pathways (SSPs)*, www.ScienceDirect.com, *Futures*, Volume 105, Pages 27-39, January 2019. https://www.sciencedirect.com/science/article/abs/pii/S001632871730349X?via%3Dihub

270 Working Group II Contribution To The Fifth Assessment Report Of The Intergovernmental Panel On Climate Change, United Nations IPCC, *Climate Change 2014 Impact, Adaptation, and Vulnerability*, www.Archive.IPCC.ch, 2014. Page accessed on January 20, 2020. https://archive.ipcc.ch/pdf/assessment-report/ar5/wg2/ar5_wgII_spm_en.pdf

271 Organization for Economic Cooperation and Development (OECD), OECD.Stat, "Aid activities targeting Global Environmental Objectives," www.Stats.OECD.org, Data extracted on January 20, 2020. https://stats.oecd.org/Index.aspx?DataSetCode=RIOMARKERS

272 Lomberg, Bjorn, "The Danger of Climate Doomsayers," www.Project-Syndicate.org, August 19, 2019. https://www.project-syndicate.org/commentary/climate-change-fear-wrong-policies-by-bjorn-lomborg-2019-08

273 The Copenhagen Consensus, The Expert Panel, "The Nobel Laureates' Guide To The Smartest Targets For The World," www.CopenhagenConsensus.com, 2016-2030. https://www.copenhagenconsensus.com/sites/default/files/post2015brochure_m.pdf

274 Lomberg. Ibid. 2019.

275 Pielke, Roger A., "Future economic damage from tropical cyclones: sensitivities to societal and climate changes," www.RoyalSocietyPublishing.org, via the *Philosophical Transactions of the Royal Society A Mathematical, Physical and Engineering Sciences*, July 30, 2017. https://royalsocietypublishing.org/doi/full/10.1098/rsta.2007.2086

276 Davis Hanson, Victor, "Energy Paradoxes Put Europe in a Precarious Position," www.Townhall.com, January 16, 2020. https://townhall.com/columnists/victor

davishanson/2020/01/16/energy-paradoxes-put-europe-in-a-precarious-position-n2559550?utm_source=thdaily&utm_medium=email&utm_campaign=nl&newsletterad=01/16/2020&bcid=f474ff7f07777dc0fd4d372af0dcd99b&recip=24783318

277 Davis Hanson. Ibid. 2020.

278 Royal, Todd, "The U.S. is in a geopolitical mess over the Nord Stream 2 pipeline," www.Cfact.org, October 25, 2018. https://www.cfact.org/2018/10/25/the-u-s-is-in-a-geopolitical-mess-over-the-nordstream-2-pipeline/

279 Davis Hanson. Ibid. 2020.

280 Van Beurden, Ben, "Two billion people do not have access to reliable electricity: this must change," www.Linkedin.com, October 18, 2018. https://www.linkedin.com/pulse/two-billion-people-do-have-access-reliable-must-ben-van-beurden/

281 Browne, Ryan, NATO report says only 7 members are meeting defense spending targets," www.CNN.com, March 14, 2019. https://www.cnn.com/2019/03/14/politics/nato-defense-spending-target/index.html

282 Foust, Michael, "Persecution of Christian Meets 'Definition of Genocide,' Report for British Gov't Says," www.ChristianHeadlines.com, July 8, 2019. https://www.christianheadlines.com/contributors/michael-foust/persecution-of-christians-meets-definition-of-genocide-report-for-british-gov-t-says.html

283 Estabrook, Rachel, "Climate Protesters Disrupt Polis' State Of The State Speech," www.CPR.org, (CPR News), January 9, 2020. https://www.cpr.org/2020/01/09/climate-protesters-disrupt-polis-state-of-the-state-speech/

284 Stevens, Pippa, "Here how the world's largest money manager is overhauling its strategy because of climate change," www.CNBC.com, January 14, 2020. https://www.cnbc.com/2020/01/14/blackrock-is-overhauling-its-strategy-to-focus-on-climate-change.html

285 Pippa. Ibid. 2020.

286 CNN Business, "DAVOS 2020," www.Edition.CNN.com, January 21, 2020. https://edition.cnn.com/videos/business/2020/01/21/microsoft-climate-change-brad-smith-davos.cnn-business

287 Starbucks Stories & News, "Starbucks announces multi-decade aspiration to become resource positive," www.Stories.Starbucks.com, January 21, 2020. https://stories.starbucks.com/stories/2020/starbucks-announces-multi-decade-aspiration-to-become-resource-positive/

288 Policelli, James, "Dangerous Fantasy: Sierra Club's 100% Wind & Solar Push Threatens US Power Supply," www.StopTheseThings.com, January 22, 2020. https://stopthesethings.com/2020/01/22/dangerous-fantas y-sierra-clubs-100-wind-solar-push-threatens-us-power-supply/

289 Tittel, Jeff, "Federal fossil-fuel subsidies hurt renewables, drive up our electric bills| Letter," www.LehighValleyLive.com, December 26, 2019. https://www.lehighvalleylive.com/opinion/2019/12/federal- fossil-fuel-subsidies-hurt-renewables-drive-up-our-electric-bills- letter.html

290 Policelli. Ibid. 2020.

291 Blohm, Robert, "The Green New Deal's Impossible Electric Grid: Renewable energy can't consistently balance power supply with demand," www.WSJ.com, February 20, 2019. https://www.wsj.com/ articles/the-green-new-deals-impossible-electric-grid-11550705997

292 Russell Mead, Walter, "All Aboard the Crazy Train: Or at least that's how populism rise feels to those at the World Economic Forum," www.WSJ.com, January 20, 2020. https://www.wsj.com/articles/al l-aboard-the-crazy-train-11579554512

293 Murphy, Peter, "Climate policies harm black and brown people," www.Cfact.org, January 20, 2020. https://www.cfact.org/2020/01/20/ climate-policies-harm-black-and-brown-communities/

294 Driessen, Paul, Wojick, David, "Reform USAID Energy Aid Policies ASAP!" www.Townhall.com, January 4, 2020. https://townhall.com/ columnists/pauldriessen/2020/01/04/reform-usaid-energy-aid-policie s-asap-n2558947

295 Bell, Larry, "'Green Energy' Capacities Are Overblown Hot Air," www.NewsMax.com, December 9, 2019. https://www.newsmax. com/larrybell/wind-fossil-petroleum-solar/2019/12/09/id/945107/

296 Simon, David, "Paul Krugman Is a Global Warming Alarmist. Don't Be Like Him," www.RealClearMarkets.com, January 16, 2020. https:// www.realclearmarkets.com/articles/2020/01/16/paul_krugman_is_ a_global_warming_alarmist_dont_be_like_him_104041.html

297 Williams, Walter, "Global Warming," www.Townhall.com, March 11, 2015. https://townhall.com/columnists/walterewilliams/2015/03 /11/global-warming-n1967847

298 Dr. Lehr, Jay, Taylor, James, "Climate models continue to project too much warming," www.Cfact.org, January 6, 2020. https://www.cfact.org/2020/01/06/climate-models-continue-t o-project-too-much-warming/

299 Editorials via Investors.com, "Don't Tell Anyone, But We Just Had Two Years Of Record-Breaking Global Cooling," www.Investors. com, May 16, 2018.h ttps://www.investors.com/politics/editorials/climate-change-global-warming -earth-cooling-media-bias/

300 Simon. Ibid. 2020.

301 Ritchie, Hannah, Roser, Max, "Natural Disasters," www. OurWorldinData.org, November 2019. Quote came from Simon. Ibid. 2020. Data verified by this database. https://ourworldindata. org/natural-disasters

302 Ritchie. Roser. Ibid. 2019.

303 Perry, Mark J., "Six facts about the non-problem of global warming," Blog Post, Carpe Diem, www.AEI.org, January 21, 2020. https://www.aei.org/carpe-diem/six-facts-about-the-non-proble m-of-global-warming/

304 Congress of the United States Congressional Budget Office, *CBO The 2019 Long-Term Budget Outlook*, June 2019. https://www.cbo.gov/ system/files/2019-06/55331-LTBO-2.pdf

305 Simon. Ibid. 2020.

306 U.S. Congressional Testimony by Kevin D. Dayaratna, PhD, "Methods and Parameters Used to Establish the Social Cost of Carbon Testimony before the Subcommittee on Environment and Oversight," Committee on Science and Technology, U.S. House of Representative, www.Docs.House.Gov, February 24, 2017. https:// docs.house.gov/meetings/SY/SY18/20170228/105632/HHRG-11 5-SY18-Wstate-DayaratnaK-20170228.pdf

307 Davis Hanson, Victor, "Government in the Shadows," www. NationalReview.com, January 21, 2020. https://www.nationalreview. com/2020/01/anti-trumpers-conduct-shadow-government/

CHAPTER FIVE: *FUTURE DEVELOPED COUNTRIES WITHOUT FOSSIL FUELS - SOCIAL CHANGES TO LIVE IN THE PRE-1900'S*

308 Petroleum Products, https://en.wikipedia.org/wiki/Petroleum product

309 Petroleum refining in the United States, https://en.wikipedia.org/wiki/ Petroleum refining in the United States

310 Wikipedia, List of Refineries, https://en.wikipedia.org/wiki/List_of_oil_refineries

311 Testimony before the Subcommittee on Environment and Oversight, February 24, 2017, https://docs.house.gov/meetings/SY/SY18/20170228/105632/HHRG-115-SY18-Wstate-DayaratnaK-20170228.pdf

312 Loris, Nicholas, The Heritage Foundation, February 8, 2019, Green New Deal Would Barely Change Earth's Temperature. Here Are the Facts, https://www.heritage.org/energy-economics/commentary/green-new-deal-would-barely-change-earths-temperature-here-are-the

313 Beisner, Calvin, November 11, 2019, Climate change: a first-world problem? https://cornwallalliance.org/2019/11/climate-change-a-first-world-problem/

314 Ben van Beurden, October 18, 2019, Two billion people do not have access to reliable electricity: this must change, https://www.linkedin.com/pulse/two-billion-people-do-have-access-reliable-must-ben-van-beurden/

315 Hasemyer, David, January 17, 2020, Inside Climate News, Fossil Fuels on Trial: Where the Major Climate Change Lawsuits Stand Today, https://insideclimatenews.org/news/04042018/climate-change-fossil-fuel-company-lawsuits-timeline-exxon-children-california-cities-attorney-general

316 Stein, Ronald, December 1, 2019, CFACT, Can we go back to the pre-fossil fuel era? https://www.cfact.org/2019/12/01/can-we-adapt-to-pre-fossil-fuel-era/

317 ELENA KOSOLAPOVA, Elena, Content Editor for Climate Change Policy and Adaptation (Russia/Netherlands), November 29, 2018, Paris Agreement Reaches 184 Ratifications, https://sdg.iisd.org/news/paris-agreement-reaches-184-ratifications/

318 Climate Home News, December 7, 2018, Which countries have not ratified the Paris climate agreement? https://www.climatechangenews.com/2018/07/12/countries-yet-ratify-paris-agreement/

319 Climate Home News, December 7, 2018, Which countries have not ratified the Paris climate agreement? https://www.climatechangenews.com/2018/07/12/countries-yet-ratify-paris-agreement/

320 Travelweek, February 17, 2017, Exactly how many planes are there in the world today? https://www.travelweek.ca/news/exactly-many-planes-world-today/

321 Hugh Morris, The Telegraph, August 16, 2017, https://www.telegraph. co.uk/travel/travel-truths/how-many-planes-are-there-in-the-world/

322 Oilprice.com, https://oilprice.com/Energy/Energy-General/ Diesel-Demand-Is-Set-To-Soar.html

323 Voelcker, John, Green, July 29, 2014, Car Reports, 1.2 billion vehicles on worlds roads now, 2 billion by 2035, https://www.greencarreports. com/news/1093560_1-2-billion-vehicles-on-worlds-roads-now-2-bi llion-by-2035-report

324 EIA, September 24, 2019, EIA projects nearly 50% increase in world energy usage by 2050, led by growth in Asia, https://www.eia.gov/ todayinenergy/detail.php?id=41433#

325 Unicef, Reduce child mortality, https://static.unicef.org/mdg/ childmortality.html

326 Unicef, Reduce child mortality, https://static.unicef.org/mdg/ childmortality.html

327 Shah, Anup, Global Issues, January 7, 2013, Poverty Facts and Stats, http://www.globalissues.org/article/26/poverty-facts-and-stats

328 DoSomething.org, 11 Facts about global poverty, https://www. dosomething.org/us/facts/11-facts-about-global-poverty

329 Coal Plants by Country https://docs.google.com/spreadsheets/ d/1I8GeKEfxPpwkQ_t0GQZx1GQm6MASclEtEtrQX3Y1nNc/ edit#gid=0

330 Erin De Santiago, How Much Fuel Does a Cruise Ship Use? https:// cruises.lovetoknow.com/wiki/How_Much_Fuel_Does_a_Cruise_ Ship_Use

331 Transport Geography, Fuel Consumption by Containership Size and Speed, https://transportgeography.org/?page_id=5955

332 https://wattsupwiththat.com/2019/09/07/there-is-no-climate-emergency/

333 Climate Change: The Facts 2017 Paperback – August 31, 2017, https:// www.amazon.com/Climate-Change-Facts-Jennifer-Marohasy/ dp/0909536031

334 Temperatures, Inconvenient Facts: The science that Al Gore doesn't want you to know, https://www.amazon.com/ Inconvenient-Facts-science-that-doesnt/dp/1545614105/ref=sr_ 1_1?crid=16BW1LLI80CRH&keywords=gregory+wrightsto ne&qid=1574352823&s=books&sprefix=gregory+wright% 2Ctoys-and-games%2C127&sr=1-1

335 Paul Driessen. Paul, September 22, 2017, CFACT, Hurricane illusions – and realities, https://www.cfact.org/2017/09/22/irma-illusions-and-realities/

336 Morano, Mark, The Politically Incorrect Guide to Climate Change (The Politically Incorrect Guides) Paperback – February 26, 2018, https://www.amazon.com/Politically-Incorrect-Climate-Change-Guides/dp/1621576760/ref=pd_bxgy_14_img_2/140-8381470-1313641? encoding=UTF8&pd_rd_i=1621576760&pd_rd_r=c1430973-b405-408b-91d4-74e396e61cdd&pd_rd_w=URIfR&pd_rd_wg=i77XX&pf

337 Driessen, Paul, September 2, 2019, More Fake Five-Alarm Crises from the IPCC September 2, 2019, https://us-issues.com/2019/09/02/more-fake-five-alarm-crises-from-the-ipcc/

338 Ranken Energy Corporation, Products made from Petroleum, https://www.ranken-energy.com/index.php/products-made-from-petroleum/

339 International Renewable Energy Agency (IRENA), Renewable Capacity Statistics 2019, https://www.irena.org/publications/2019/Mar/Renewable-Capacity-Statistics-2019#targetText=March%202019&targetText=Renewable%20power%20generation%20capacity%20is,end%20of%20the%20calendar%20year.

340 Chatsko, Maxx, Jun 4, 2018, Big Oil Is Investing Billions in Renewable Energy, https://www.fool.com/investing/2018/06/04/big-oil-is-investing-billions-in-renewable-energy.aspx

341 Shellenberger, Michael, Forbes Magazine, May 15, 2018, Solar And Wind Lock-In Fossil Fuels, And That Makes Saving The Climate Harder And More Expensive, https://www.forbes.com/sites/michaelshellenberger/2018/05/15/solar-and-wind-lock-in-fossil-fuels-that-makes-saving-the-climate-harder-slower-more-expensive/#2417a24921d4

342 Voelcker, John, Green, July 29, 2014, Car Reports, 1.2 billion vehicles on worlds roads now, 2 billion by 2035, https://www.greencarreports.com/news/1093560_1-2-billion-vehicles-on-worlds-roads-now-2-billion-by-2035-report

343 Stein, Ronald, New Geography, September 11, 2019, GREEN TECHNOLOGY'S DARK SIDE, https://www.newgeography.com/content/006406-green-technologys-dark-side

344 Amnesty Organization, November 15, 2017, Industry giants fail to tackle child labour allegations in cobalt battery supply chains, https://www.amnesty.org/en/latest/news/2017/11/industry-giants-fail-to-tackle-child-labour-allegations-in-cobalt-battery-supply-chains/

345 SB657, The California Transparency in Supply Chains Act, https:// oag.ca.gov/SB657

346 H.R.4842, Business Supply Chain Transparency on Trafficking and Slavery Act of 2014, https://www.congress.gov/bill/113th-congress/ house-bill/4842

347 Williams, Jason, April 21, 2017, Wealth Daily, Elon Musk's Dirty Secret, https://www.wealthdaily.com/articles/elon-musk-s-dirty-secret/90282

348 Kolodny, Lora, August 20, 2019, Walmart sues Tesla over solar panel fires at seven stores, https://www.cnbc.com/2019/08/20/ walmart-sues-tesla-over-solar-panel-fires-at-seven-stores.html

349 Kolodny, Lora, CNBC, August 24, 2019,Tesla solar panels reportedly caught fire at an Amazon warehouse in 2018, https://www.msn. com/en-us/finance/companies/tesla-solar-panels-reportedly-caugh t-fire-at-an-amazon-warehouse-in-2018/ar-AAGf5EV

350 Axelrod, Tal, The Hill, December 31, 2019, Merkel vows Germany will do 'everything humanly possible' to fight climate change, https:// thehill.com/policy/energy-environment/476416-merkel-vows-german y-will-do-everything-humanly-possible-to-fight

351 Reed, Chris, CalWatchDog.com, November 25, 2019,, Gov. Newsom suspends new fracking permits in latest attempt to reduce greenhouse gas emissions, https://calwatchdog.com/2019/11/25/ gov-newsom-suspends-new-fracking-permits-in-latest-attempt-to-red uce-greenhouse-gas-emissions/

352 California Energy Commission, Oil Supply Sources to California Refineries, https://ww2.energy.ca.gov/almanac/petroleum_data/ statistics/crude_oil_receipts.html

353 Grady, John, USNI News, November 28, 2019, U.S. Shoulders Steep Price to Protect Merchant Ships in Strait Of Hormuz, https://news.usni.org/2019/11/29/report-u-s-shoulders-stee p-price-to-protect-merchant-ships-in-strait-of-hormuz?utm_ source=USNI+News&utm campaign=6d765a22f0- USNI NEWS DAILY&utm medium=email&utm term=0_0dd4a1450b-6d765a22f0-233739661&ct=t(USNI_NEWS_ DAILY)&mc_cid=6d765a22f0&mc_eid=9408dbff30

354 California Energy Commission, Oil Supply Sources to California Refineries, https://ww2.energy.ca.gov/almanac/petroleum_data/ statistics/crude_oil_receipts.html

355 Clear Energy Alliance, July 14, 2019, California Illusion video, https:// www.youtube.com/watch?v=m9I0NLoAXNQ

356 Ferrar, Kyle, Fractracker Alliance, July 2, 2019, Impact of a 2,500' Oil and Gas Well Setback in California, https://www.fractracker.org/2019/07/impact-of-a-2500-oil-and-gas-well-setback-in-california/

357 SB657, The California Transparency in Supply Chains Act, https://oag.ca.gov/SB657

358 H.R.4842, Business Supply Chain Transparency on Trafficking and Slavery Act of 2014, https://www.congress.gov/bill/113th-congress/house-bill/4842

359 Stein, Ronald, CFACT, September 24, 2018, America is following Germany's failed climate goals, https://www.cfact.org/2018/09/24/america-is-following-germanys-failed-climate-goals/

360 Stein, Ronald, Fox & Hounds, May 29, 2019, Australia's voters reject environmental fantasies, http://www.foxandhoundsdaily.com/2019/05/australias-voters-reject-environmental-fantasies/

361 Paul Rogers and Katy Murphy, September 10, 2018, The Mercury News, California mandates 100 percent clean energy by 2045, https://www.mercurynews.com/2018/09/10/california-mandates-100-percent-clean-energy-by-2045/

362 EIA, California Price Differences from U.S. average, https://www.eia.gov/state/?sid=CA#tabs-5

363 Dickerson. Kelly, Business Insider, May 12, 2015, The world's lust for new technology is creating a 'hell on Earth' in Inner Mongolia, https://www.businessinsider.com/the-worlds-tech-waste-lake-in-mongolia-2015-5

364 Bell, Terrence, November 20, 2019, The World's Biggest Cobalt Refiners, China, Finland, Belgium and Norway are among the top refiners, https://www.thebalance.com/the-biggest-cobalt-producers-2339726

365 Jones, Barbara, August 6, 2017, Child miners aged four living a hell on Earth so YOU can drive an electric car, https://www.dailymail.co.uk/news/article-4764208/Child-miners-aged-four-living-hell-Earth.html

366 Tim Maughan, BBC, April 2, 2015, Hidden in an unknown corner of Inner Mongolia is a toxic, nightmarish lake created by our thirst for smartphones, consumer gadgets and green tech, https://www.bbc.com/future/article/20150402-the-worst-place-on-earth

CHAPTER SIX: *BILLIONS WITHOUT BASIC ELECTRICITY - AFFORDABILITY WILL BE THE FUTURE*

367 U.S. Energy Information Administration, Independent Statistics & Analysis, Today in Energy, "EIA forecasts U.S. crude oil production will keep growing through 2021, but more slowly," www.EIA.gov, January 27, 2020. https://www.eia.gov/todayinenergy/detail.php?id=42615

368 Rystad Energy, Press Release, "New Mexico Flirts With 1 Million BPD Oil Threshold," www.RystadEnergy.com, January 21, 2020. https://www.rystadenergy.com/newsevents/news/press-releases/new-mexico-flirts-with-1-million-bpd-oil-threshold/

369 Nuclear Energy Institute (NEI) Media Team, Statement, Advanced Nuclear, "Congress Funds Nuclear Carbon-Free Energy at Historic Levels," www.NEI.org, December 19, 2019. https://www.nei.org/news/2019/2019-congressional-nuclear-spending-bill

370 Nuclear Energy Institute (NEI), Climate, "Nuclear energy provides more than 55 percent of America's carbon-free electricity," www.NEI.org, 2020. Page accessed January 28, 2020. https://www.nei.org/advantages/climate

371 van Beurden, Ben, "Two billion people do not have access to reliable electricity: this must change," www.Linkedin.com, October 18, 2018. https://www.linkedin.com/pulse/two-billion-people-do-have-access-reliable-must-ben-van-beurden/

372 Parke, Phoebe, "Why are 600 million Africans still without power?" www.CNN.com, April 1, 2016. https://www.cnn.com/2016/04/01/africa/africa-state-of-electricity-feat/index.html

373 Clear Energy Alliance, "BERNing Down America," www.YouTube.com, January 20, 2020. https://www.youtube.com/watch?v=TNtOpX3I9jg

374 Stromsta, Karl-Erik, "BlackRock Targets Storage With New Multibillion-Dollar Renewables Fund," www.GreenTechMedia.com, January 27, 2020. https://www.greentechmedia.com/articles/read/blackrock-targets-storage-with-new-multi-billion-dollar-renewables-fund?utm_source=newsletter&utm_medium=email&utm_campaign=newsletter_axiosgenerate&stream=top

375 Admin, "Forget Paris: 1,600 New Coal-fired Power Plants are Planned or Under Construction in 62 Countries," www.SaltBushClub.com,

January 21, 2019. https://saltbushclub.com/2019/01/21/1600-ne
w-coal-power-plants-worldwide/

376 Dohmen, Frank, Jung, Alexander, Schultz, Stefan, Traufetter, Gerald,
Climate Stasis, "German Failure on the Road to a Renewable Future,"
www.Spiegel.de, (Spiegel International), May 13, 2019. https://www.
spiegel.de/international/germany/german-failure-on-the-road-to-a-re
newable-future-a-1266586.html

377 Moran, Alan, "Renewables rent-seekers aren't interested in bushfire
prevention – or cheap efficient energy," www.Spectator.com.
au, January 20, 2020. https://www.spectator.com.au/2020/01/
renewables-rent-seekers-arent-interested-in-bushfire-pre
vention-or-cheap-efficient-energy/

378 MacLellan, Kylie, "Electric dream: Britain to ban new petrol and hybrid
cars from 2035," www.Reuters.com, February 3, 2020. https://www.
reuters.com/article/us-climate-change-accord/electric-future-britain-t
o-ban-new-petrol-and-hybrid-cars-from-2035-idUSKBN1ZX2RY

379 Booth, David, "Motor Mouth: More inconvenient truths on
banning gas engines," www.Driving.CA, October 6, 2017. https://
driving.ca/auto-news/news/motor-mouth-more-inconvenien
t-truths-on-banning-gas-engines

380 Booth. Ibid. 2017.

381 Booth. Ibid. 2017. Entire paragraph.

382 Renewable Energy Foundation (REF), REF Blog, "A Decade of
Constraint Payments," www.Ref.Org.Uk, December 30, 2019. https://
www.ref.org.uk/ref-blog/354-a-decade-of-constraint-payments

383 EurekaAlert!, American Psychological Association, "Majority of US
adults believe climate change is most important issue today," www.
EurekaAlert.org, February 6, 2020. https://www.eurekalert.org/pub
releases/2020-02/apa-mou020620.php#.XjxgnT10CRk.gmail

384 Kyiv Post, Ukraine's Global Voice, via Financial Times (FT),
"Financial Times: Russia to go it alone on construction of Nord
Stream 2 pipeline," www.KYIVPost.com, January 28, 2020. https://
www.kyivpost.com/ukraine-politics/financial-times-russia-to-go-it
-alone-on-construction-of-nord-stream-2-pipeline.html

385 Royal, Todd, "A Century of Russia's Weaponization of Energy," www.
ModernDiplomacy.eu, October 14, 2019. https://moderndiplomacy.
eu/2019/10/14/a-century-of-russias-weaponization-of-energy/

386 Constable, John, "The burden of proof rests on Mark Carney,
and he hasn't made his case against fossil fuels," www.Business.

FinancialPost.com, January 10, 2020. https://business.financialpost.
com/opinion/the-burden-of-proof-rests-on-mark-carney-and-he-has
nt-made-his-case-against-fossil-fuels

387 Mills, Mark P., "The "New Energy Economy": An Exercise in Magical Thinking," Energy & Environment: Technology/Infrastructure, www.Manhattan-Institute.org, March 26, 2019.

388 Baker, David R., Wade, Will, Thornhill, James, Climate Changed section, "Sometimes, a Greener Grid Means a 40,000% Spike in Power Prices," www.Bloomberg.com, August 25, 2019. https://www.bloomberg.com/news/articles/2019-08-26/ sometimes-a-greener-grid-means-a-40-000-spike-in-power-prices

389 Bordoni, Linda, "Christians persecuted in Nigeria amid deafening silence," www.VaticanNews.va, January 28, 2020.

390 StopTheseThings.com, "Full-Steam Ahead: China & Japan Snub Intermittent Wind & Solar to Build Hundreds of New-Age Coal-Fired Plants," www.StopTheseThings.com, October 1, 2018. https://stopthesethings.com/2018/10/01/full-steam-ahead-china-japan-snub-intermittent-wind-solar-to-build-hundreds-of-n ew-age-coal-fired-plants/

391 International Energy Agency (IEA), World Energy Outlook 2019, (IEA publications, Paris, France), November 2019. https://www.iea. org/reports/world-energy-outlook-2019

392 Mansfield, Harvey C., "Self-Interest Rightly Understood," Journal Article, Political Theory, Vol. 23. No. 1, Pages 48-66, www.Jstor. org, February 1995. https://www.jstor.org/stable/192173?seq=1

393 StopTheseThings.com. Ibid. 2018 & 2020.

394 Doshi, Tilak, "In Coal We Trust: The Need For Coal Power In Asia," www.Forbes.com, June 7, 2019. https://www.forbes. com/sites/tilakdoshi/2019/06/07/in-coal-we-trust-the-need-fo r-coal-power-in-asia/#306c7811222c

395 Lomberg, Bjorn, TED talk, Ideas worth spreading, "Global priorities bigger than climate change," www.Ted.com, February 2005. Mr. Lomberg to my knowledge has never taken back what he said about climate change though he believes the issue needs to be addressed. https://www.ted.com/talks/bjorn_lomborg_global_priorities_bigger_ than_climate_change?language=en

396 International Energy Agency World Energy Outlook. Ibid. 2019.

397 International Energy Agency World Energy Outlook. Ibid. 2019.

398 Krebs, Mark, Tanton, Tom, "Problems of Industrial Electrification (forced decarbonization on the firing line,") www. MasterResource.org, January 9, 2020. https://www.masterresource. org/deep-decarbonization-electricity-for-gas-mandates/ problems-of-industrial-electrification/

399 Krebs. Tanton. Ibid. 2020.

400 Krebs, Mark, "Costing the Green New Deal and "Deep Decarbonization": Some Clarifications," www.MasterResource.org, July 11, 2019. https://www.masterresource.org/krebs-mark/costin g-the-green-new-deal-and-deep-decarbonization-some-clarifications/

401 Mills. Ibid. 2019.

402 Krebs. Tanton. Ibid. 2020.

403 Bradley Jr., Robert, "Investors Confront Tesla's Energy Fantasy," www.Forbes.com, August 24, 2016. https:// www.forbes.com/sites/robertbradley/2016/08/24/ investors-confront-teslas-energy-fantasy/#5744d0cc16e0

404 Constable. Ibid. 2020.

405 International Energy Agency (IEA), "Data and Statistics," Explore energy data by category, indicator, country or region. www.IEA. org, 2020. Page accessed January 30, 2020. https://www.iea. org/data-and-statistics?country=WORLD&fuel=Energy%20 supply&indicator=Total%20primary%20energy%20supply%20 (TPES)%20by%20source

406 CNBC, "How fracking changed America forever," www.MSN.com (MSN Money section), January 7, 2020. https://www.msn.com/en-us/ money/markets/how-fracking-changed-america-forever/ar-BBYIpw0

407 U.S. Energy Information Administration, Independent Statistics & Analysis, Analysis & Projections, *International Energy Outlook 2019*, (U.S. Department of Energy, Washington, D.C.), Release date: September 24, 2019. Next release date: September 2020. https:// www.eia.gov/outlooks/ieo/

408 Kotkin, Joel, Cox, Wendell, "California's Inept Central Planners," www. NewGeography.com, January 13, 2020. https://www.newgeography. com/content/006526-californias-inept-central-planners

409 FuelsEurope, "Where is crude oil used? What products and applications are made from oil?" www.YouTube.com, June 6, 2014. https://www.youtube.com/watch?v=fKGgxFnoGK8

410 Kotkin. Cox. Ibid. 2020.

411 Jackson, Kerry, "OTHER VOICES: California Green New Deal embraces far left policy wish list under guise of saving the planet," www.Bakersfield.com, January 12, 2020.

412 From Barnabas Fund Contacts, "Full scale jihad" unfolds in Nigeria as Fulani kill thirteen Christians amid ceaseless Boko Haram bloodshed," www.BarnabusFund.org, January 28, 2020. https://barnabasfund.org/en/news/"full-scale-jihad"-unfolds-in-nigeria-as-fulani-kill-thirteen-christians-amid-ceaseless-boko

413 I & I Editorial Board, "America's Next Housing Shortage, Brought To You By The Democrats," www.IssuesInsights.com, January 31, 2020. https://issuesinsights.com/2020/01/31/americas-next-housing-shortage-brought-to-you-by-the-democrats/

414 I & I Editorial Board. Ibid. 2020.

415 Kotkin, Joel, Cox, Wendell, Eye On The News, Economy, finance, and budgets; California, "In Defense of Houses: Single-family homes are the backbone of American aspiration – so why do so many oppose them?" www.City-Journal.org, July 16, 2019. https://www.city-journal.org/single-family-housing-opposition

416 Davis Hanson, Victor, "The Cult of West-Shaming," www.NationalReview.com, January 30, 2020. https://www.nationalreview.com/2020/01/woke-elites-criticize-west-silent-on-china-iran/

417 Hanson. Ibid. 2020.

418 Folley, Aris, "Greta Thunberg nominated for Nobel Peace Prize," www.TheHill.com, February 3, 2020. https://thehill.com/blogs/in-the-know/in-the-know/481135-greta-thunberg-nominated-for-nobel-peace-prize

419 Kirby, Felix, "Teenage Climate-Change Protestors Have No Idea Why They're Protesting," www.Quilette.com, April 25, 2019. https://quillette.com/2019/04/25/teenage-climate-change-protestors-have-no-idea-what-theyre-protesting/

420 Wikipedia, "Montreal Protocol," www.Wikipedia.com, Last edited December 17, 2019. Page accessed on February 5, 2020. https://en.wikipedia.org/wiki/Montreal_Protocol

421 Dr. Lehr, Jay, "Luke-warming: The climate campaign's cottage industry," www.Cfact.org, February 4, 2020. https://www.cfact.org/2020/02/04/luke-warming-the-climate-campaigns-cottage-industry/

422 Dr. Lehr, Jay. Ibid. 2020.

423 Dr. Lehr, Jay. Ibid. 2020.

424 Hanson. Ibid. 2020.

425 U.S. Energy Information Administration (EIA), Independent Statistics & Analysis, Today In Energy, "U.S. crude oil and natural gas production increased in 2018, with 10% fewer wells," www.EIA.gov, February 3, 2020. https://www.eia.gov/todayinenergy/detail.php?id=42715

426 StopTheseThings.com, "Setting It Straight: Nuclear Power Sets the Gold Standard For Safe & Reliable Power Generation," www.StopTheseThings.com, February 3, 2020. https://stopthesethings.com/2020/02/03/setting-it-straight-nuclear-power-sets-the-gold-standard-for-safe-reliable-power-generation/comment-page-1/

427 Lambermont, Paige, "Three Mile Island and the Exaggerated Risk of Nuclear Power," www.Fee.org, (Foundation for Economic Education), January 14, 2020. https://fee.org/articles/three-mile-island-and-the-exaggerated-risk-of-nuclear-power/

428 Douglas, Holger (Translated/edited by Pierre Gosselin), "Germany's Energiewende (transition to green energies) is driving up prices," www.NoTricksZone.com, via www.StopTheseThings.com, January 26, 2020. https://stopthesethings.com/2020/02/02/renewable-energy-transition-wind-solar-obsession-leaves-germans-suffering-the-worlds-highest-power-prices/

429 World Nuclear News, "Viewpoint: There is no Holy Grail of energy," www.World-Nuclear-News.org, December 17, 2018. https://www.world-nuclear-news.org/Articles/Viewpoint-There-is-no-Holy-Grail-of-energy

430 Perry, Mark J., Blog Post, "Winston Churchill speaking on the dangers of socialism in 1945, just as relevant today," Carpe Diem, www.AEI.org, (American Enterprise Institute), January 29. 2020. https://www.aei.org/carpe-diem/winston-churchill-speaking-on-the-dangers-of-socialism-in-1945-just-as-relevant-today/

431 Citizen Budget Commission (CBC), *Getting Greener: Cost-Effective Options for Achieving New York State's Greenhouse Gas Goals*, Report, State Budget, www.CBCNY.org, December 9, 2019. https://cbcny.org/research/getting-greener

432 Caiazza, Roger, "New York Energy Bill from Cuomo to Consumers Could Top $47 Billion!" www.NaturalGasNow.com, January 8, 2020. https://naturalgasnow.org/new-york-energy-bill-from-cuomo-to-consumers-could-top-47-billion/

433 CBC. Ibid. 2020.

434 CBC. Ibid. 2020.

435 Caiazza. Ibid. 2020.

436 Caiazza. Ibid. 2020.

437 Stein, Ronald, Royal, Todd, *Energy Made Easy: Helping Citizens Become Energy Literate*, (Xlibris Publishing, Bloomington, IN.), Chapter Six: Renewable Electricity, August 12, 2019. https://www.amazon.com/dp/B07WQ2KZG3/ref=rdr_kindle_ext_tmb

438 Caiazza. Ibid. 2020.

439 Caiazza. Ibid. 2020.

440 CBC Forecasted 2040 Capacity Resources to Meet CLCPA Goals During January 3-4 2018 Winter Peak, (CBC Forecasted 2040 CLCPA Goals), Spreadsheet Table/Chart. Page accessed February 6, 2020. https://pragmaticenvironmentalistofnewyork.files.wordpress.com/2020/01/cbc-forecasted-2040-capacity-resources-to-meet-clcpa-goals-during-january-3-4-2018-winter-peak.pdf

441 CBC Forecasted 2040 CLCPA Goals. Ibid. 2018.

442 Fu, Ran, Remo, Timothy, Margolis, Robert, National Renewable Energy Laboratory, U.S. Department of Energy, Office of Energy Efficiency & Renewable Energy, *2018 U.S. Utility-Scale Photovoltaics-Plus-Energy Storage System Costs Benchmark*, (U.S. Department of Energy, Washington, D.C.), November 2018. https://www.nrel.gov/docs/fy19osti/71714.pdf

443 Dears, Donn, "Four Minutes for $150 million," www.DDears.com, January 15, 2019. https://stopthesethings.com/2019/02/09/insane-storage-cost-mean-batteries-no-solution-to-chaotic-wind-power-delivery/

444 Davis Hanson, Victor, "Progressive Petards," www.AMGreatness.com, January 19, 2020. https://amgreatness.com/2020/01/19/progressive-petards/

CHAPTER SEVEN: *CLIMATE ALARMISTS WHO BENEFIT FROM THE ALL-ELECTRIC NARRATIVE*

445 The United Nations Intergovernmental Panel on Climate Change (IPCC), "Summary for Policymakers of IPCC Special Report on Global Warming of 1.5 Celsius approved by governments," www.IPCC.Ch, October 8, 2018. https://www.ipcc.ch/2018/10/08/summary-for-policymakers-of-ipcc-special-report-on-global-warming-of-1-5c-approved-by-governments/

446 Pielke, Roger, "The World Is Not Going To Halve Carbon Emissions By 2030, So Now What?" www.Forbes.com, October 27, 2019. https://www.forbes.com/sites/rogerpielke/2019/10/27/the-world-is-not-going-to-reduce-carbon-dioxide-emissions-by-50-by-2030-now-what/#51acc8723794

447 Pielke, Roger, *The Climate Fix*, (Basic Books, New York, NY), Entire Book, December 6, 2011. https://www.amazon.com/Climate-Fix-Roger-Pielke-Jr/dp/0465025196/ref=tmm_pap_swatch_0?_encoding=UTF8&qid=&sr=

448 British Petroleum (BP), *BP Statistical Review of World Energy 2019: an unsustainable path*, www.BP.com, June 11, 2019. https://www.bp.com/en/global/corporate/news-and-insights/press-releases/bp-statistical-review-of-world-energy-2019.html

449 BP. *BP Statistical Review of World Energy 2019*. Ibid. 2019.

450 Pielke. Ibid. 2019.

451 Loris, Nicolas, "Green New Deal Would Barely Change Earth's Temperature. Here Are the Facts," www.Heritage.org, February 8, 2019. https://www.heritage.org/energy-economics/commentary/green-new-deal-would-barely-change-earths-temperature-here-are-the (This source is used since it gives an overall global picture, U.S. focus, and links to Congressional testimony – that was sourced in previous chapters – about the U.S. shutting down, and global emissions continuing their upward, global trend.)

452 Pielke. Ibid. 2019.

453 Bloomberg News, Climate Changed, "China Set for Massive Coal Expansion in Threat to Climate Goals," www.Bloomberg.com, November 20, 2019. https://www.bloomberg.com/news/articles/2019-11-20/china-set-for-massive-coal-expansion-in-threat-to-climate-goals

454 Pielke, Roger, "Net-Zero Carbon Dioxide Emissions By 2050 Requires A New Nuclear Power Plant Every Day," www.Forbes.com, September 30, 2019. https://www.forbes.com/sites/rogerpielke/2019/09/30/net-zero-carbon-dioxide-emissions-by-2050-requires-a-new-nuclear-power-plant-every-day/#5134d72635f7

455 Pielke, Roger, "Democrat Climate Policies Are Ambitious But Fail The Reality Test," www.Forbes.com, September 9, 2019. https://www.forbes.com/sites/rogerpielke/2019/09/09/democratic-candidates-climate-policy-commitments-are-incredibly-ambitious-but-fail-a-reality-test/#7d1c6fc76faa

456 Dr. Whitehouse, David, "Are Ocean Currents Speeding Up... Or Are They Slowing Down? Nobody Knows," www.TheGWPF. com (The Global Warming Policy Forum), February 10, 2020. https://www.thegwpf.com/are-ocean-currents-speeding-up-or-are-they-slowing-down/

457 Martin, Mark, "Another 'Little Ice Age' Coming? NASA's Solar Activity Forecast Might Surprise You," www1.CBN.com, January 30, 2020. https://www1.cbn.com/cbnnews/world/2020/january/anothe r-little-ice-age-in-the-future-nasas-solar-activity-forecast-might-surpr ise-you

458 Kent, Simon, "BBC To Partner With Greta Thunberg For (Another) TV Climate Series," www.Breitbart.com, February 10, 2020. https:// www.breitbart.com/entertainment/2020/02/10/bbc-to-partner-wit h-greta-thunberg-for-another-tv-climate-series/

459 Meritnation, "If the Earth was 24 hours old, how old would humankind be?" www.Meritnation.com, April 10, 2015. https:// www.meritnation.com/blog/history-of-earth-on-24-hour-clock/

460 Dr. Lehr, Jay, Taylor, James, "Climate models continue to project too much warming," www.Cfact.org, January 6, 2020. https://www.cfact.org/2020/01/06/climate-models-continue-t o-project-too-much-warming/

461 Orlowski, Andrew, "Top boffin Freeman Dyson on climate change, interstellar travel, fusion, and more," www.TheRegister.Co.Uk, October 11, 2015. https://www.theregister.co.uk/2015/10/11/ freeman_dyson_interview/

462 Allon, Cap, "Bad News Alarmists – Official Data Reveals Arctic Sea Ice Is Once Again Growing," www.Electroverse.net, January 29, 2020. https://electroverse.net/official-data-reveals-arctic-se a-ice-is-growing-again/

463 International Energy Agency (IEA), "Global CO2 emissions in 2019," Data Release: Global energy-related CO2 emissions flattened in 2019 at around 33 gigatonnes (Gt), following two years of increases," www.IEA.org, February 11, 2020. https://www.iea.org/ articles/global-co2-emissions-in-2019

464 U.S. Energy Information Administration (EIA), Independent Statistics & Analysis, Today in Energy, "Despite the U.S. becoming a net petroleum exporter, most regions are still net importers," www.EIA. gov, February 6, 2020. https://www.eia.gov/todayinenergy/detail. php?id=42735

465 Dears, Donn, "Have You Heard of MOPR?" Power for USA, Energy Facts From Oil to Electricity, www.DDears.com, January 24, 2020. https://ddears.com/2020/01/24/have-you-heard-of-mopr/

466 Nace, Trevor, "NASA Says Earth Is Greener Today Than 20 Years Ago Thanks To China, India," www.Forbes.com, February 28, 2019. https://www.forbes.com/sites/trevornace/2019/02/28/nasa-says-earth-is-greener-today-than-20-years-ago-thanks-to-china-india/?utm_source=FACEBOOK&utm_medium=social&utm_term=Valerie%2F&fbclid=IwAR3-N4dca_Q-mp1FPXhWUlg_eWg6trfzvGA7aRiKXjaOnFxA5gjORR44HQo#61b675e76e13

467 Hirth, Lion, "The market value of variable renewables The effect of solar wind power variability on their relative price," www.Neon-Energie.de, Energy Economics, February 19, 2013. https://www.neon-energie.de/Hirth-2013-Market-Value-Renewables-Solar-Wind-Power-Variability-Price.pdf

468 Rogers, Norman, "The Climate-Industrial Complex," www. AmericanThinker.com, September 27, 2013. https://www.americanthinker.com/articles/2013/09/the_climate-industrial_complex.html

469 U.S. Energy Information Administration (EIA), Independent Statistics & Analysis, "Oil: crude and petroleum products explained," www.EIA.gov, Last updated: May 23, 2019. https://www.eia.gov/energyexplained/oil-and-petroleum-products/

470 Liberal Forum Archives, "The 140 Year Failed History of "GoreBull Warming" and Ice Ages Doom," www.LiberalForum.org, November 25, 2018. https://www.liberalforum.org/topic/243703-the-140-year-failed-history-of-gorebull-warming-and-ice-ages-doom-in-other-words-the-sky-is-falling-mantra-has-been-going-on-for-decades-and-no-one-really-gives-a-flip-because-of-all-the-lies/

471 Mahler, Joyce, "Global Cooling Will Kills Us All. No, Wait Global Warming Will Kill Us!" www.Whatsorb.com, October 23, 2019. https://www.whatsorb.com/climate/global-cooling-will-kills-us-all-no-wait-global-warming-will-kill-us

472 Elks, Sonia, "Children suffering eco-anxiety over climate change, psychologists say," www.Reuters.com, September 19, 2019. https://www.reuters.com/article/us-britain-climate-children/children-suffering-eco-anxiety-over-climate-change-say-psychologists-idUSKBN1W42CF

473 Elk. Ibid. 2019.

474 McKenzie, Barbara, "The Globalism of Climate: How Faux Environmental Concern Hides Desire to Rule the World," www. Stovouno.org, February 23, 2019. https://stovouno.org/2019/02/23/globalism-of-climate-how-faux-environmental-concern-hides-desire-to-rule-the-world/

475 Windows on the World, "Fake Protest: Extinction Rebellion," www.WindowsontheWorld.net,

476 Shellenberger, Michael, "Why Climate Alarmism Hurts Us All," www.Forbes.com, December 4, 2019. https://www.forbes.com/sites/michaelshellenberger/2019/12/04/why-climate-alarmism-hurts-us-all/#33b6a81836d8

477 Wikipedia.com, "Useful Idiot," www.Wikipedia.com, February 13, 2020. https://en.wikipedia.org/wiki/Useful_idiot

478 Kotkin, Joel, "Demographic undestiny," www.OCRegister.com, January 25, 2020. https://www.ocregister.com/2020/01/25/demographic-undestiny/

479 Burnett, H. Sterling, Driessen, Paul, "Climategate: Lessons on the 10th Anniversary (Guest; Paul Driessen), www.Heartland.org, November 21, 2019. https://www.heartland.org/multimedia/podcasts/climategate-lessons-on-the-10th-anniversary-guest-paul-driessen

480 Davis Jr., Samuel, "Stop Playing Games With America's Energy Future," www.RealClearEnergy.com, February 11, 2020. https://www.realclearenergy.org/articles/2020/02/11/stop_playing_games_with_americas_energy_future_483924.html

481 Meyer, Richard, *Dispatching Direct Use: Achieving Greenhouse Gas Reductions With Natural Gas In Homes And Businesses*, Manager Energy Analysis & Standards, American Gas Association, www.AGA.org, November 16, 2015. https://www.aga.org/sites/default/files/dispatching_direct_use_-_achieving_greenhouse_gas_reductions_the_use_of_natural_gas_in_homes_and_businesses.pdf

482 Legates, David, "It's Not About the Climate – It Never Was," www.Townhall.com, March 1, 2019. https://townhall.com/columnists/davidlegates/2019/03/01/its-not-about-the-climateit-never-was-n2542428

483 Legates. Ibid. 2019.

484 Legates. Ibid. 2019.

485 Legates. Ibid. 2019.

486 Legates. Ibid. 2019.

487 Legates. Ibid. 2019. Entire paragraph.

488 Legates. Ibid. 2019.
489 Kruta, Virginia, "'I Was Laughing So Hard I Nearly Cried': WSJ's Kimberley Strassel Can't Get Over Green New Deal," www.DailyCaller.com, February 7, 2019. https://dailycaller.com/2019/02/07/wsj-kimberley-strassel-green-new-deal/
490 Podesta, John, Goldfluss, Christy, Higgins, Trevor, Bhattacharyya, Bidisha, Yu, Alan, "A 100 Percent Clean Future," www.AmericanProgress.org, October 10, 2019. https://www.americanprogress.org/issues/green/reports/2019/10/10/475605/100-percent-clean-future/
491 Thomsen, Michael, "The US cut its CO2 emissions more than any other country in the world in 2019, helping to keep total global emissions from growing past 2018's record-breaking high," www.DailyMail.co.uk, February 11, 2020. https://www.dailymail.co.uk/sciencetech/article-7992751/2019-global-CO2-emissions-match-time-high-2018-cuts-emissions.html
492 Perry, Mark J., "Six facts about the non-problem of global warming," www.AEI.org, January 20, 2020. https://www.aei.org/carpe-diem/six-facts-about-the-non-problem-of-global-warming/
493 Jayaraj, Vijay, "No Roads and No Electricity: Why Fossil Fuels are Indispensible for Development," www.CornwallAlliance.org, July 26, 2019. https://cornwallalliance.org/2019/07/no-roads-and-no-electricity-why-fossil-fuels-are-indispensable-for-development/
494 Davis Hanson, Victor, *The Second World Wars: How The First Global Conflict Was Fought and Won*, (Basic Books, New York, NY), Chapter 19: The Dead. October 2017.
495 Jeffrey, Lauren, "An Open Letter To Extinction Rebellion," www.YouTube.com, October 21, 2019. https://www.youtube.com/watch?v=iyYPLkWV3l0
496 Shellenberger, Michael, "Why Apocalyptic Claims About Climate Change Are Wrong," www.Forbes.com, November 25, 2019. https://www.forbes.com/sites/michaelshellenberger/2019/11/25/why-everything-they-say-about-climate-change-is-wrong/#613caf8c12d6
497 McKibben, Bill, His opinion in an excerpt from Twitter, www.Twitter.com, November 23, 2019. https://twitter.com/billmckibben/status/1198375629638586373?s=20
498 Ahmed, Nafeez, "The Collapse of Civilization May Have Already Begun," www.Vice.com, November 22, 2019. https://www.

vice.com/en us/article/8xwygg/the-collapse-of-civilization-ma
y-have-already-begun

499 Cummings, William, "'The world is going to end in 12 years if we don't address climate change,' Ocasio-Cortez," www.Forbes.com, January 22, 2019. https://www. usatoday.com/story/news/politics/onpolitics/2019/01/22/ ocasio-cortez-climate-change-alarm/2642481002/

500 The Intergovernmental Panel on Climate Change (IPCC), Home page with access to all climate reports, www.IPCC.ch, Page accessed February 14, 2020. https://www.ipcc.ch

501 Ritchie, Hannah, Roser, Max, "Natural Disasters," www. OurWorldinData.org, First published in 2014, it was last revised in November 2019. https://ourworldindata.org/natural-disasters

502 Food and Agriculture Organization of the United Nations (FAO), FAO. 2018. *The future of food and agriculture – Alternative pathways to 2050*. Rome. 224 pp. License: CC BY-NC-SA 3.0 IGO. http://www. fao.org/3/I8429EN/i8429en.pdf

503 Asseng, S., Wert, F., Martre, P., et al. (50 more authors) (2015) Rising temperature reduce global wheat production. Nature Climate Change, 5 (2). Pp. 143-147. ISSN 1758-678X. http://eprints.whiterose. ac.uk/85540/1/Main Asseng 2014-9-22.pdf

504 Shellenberger. Ibid. 2019.

505 Intergovernmental Panel on Climate Change (IPCC), *Climate Change 2014, Synthesis Report*, A Report of the Intergovernmental Panel on Climate Change. First published 2015. https://www.ipcc.ch/site/ assets/uploads/2018/05/SYR AR5 FINAL full wcover.pdf

506 Nordhaus, William, "Integrated Assessment Models of Climate Change," www.Data.Nber.org, (The U.S. National Bureau of Economic Research), NBER Reporter 2017 Number 3. https://data. nber.org/reporter/2017number3/nordhaus.html

507 Shellenberger. Ibid. 2019.

508 Deacon, Ben, Timms, Penny, "More than climate change driving Queensland fires, explains climatologists," www.ABC.net.au, September 9, 2019. https://www.abc.net.au/news/2019-09-10/mor e-than-climate-change-driving-queensland-fires/11493950

509 Asmelash, Leah, "Australia's indigenous people have a solution for the country's bushfires. And it's been around for 50,000 years," www.amp.CNN.com, January 12, 2020. https://amp.cnn.com/ cnn/2020/01/12/world/aboriginal-australia-fire-trnd/index.html

510 Syphard, Alexandria D., Keeley, Jon E., Pfaff, Anne H., Ferschweiler, Ken, "Human presence diminishes the importance of climate in driving fire activity across the United States," Conservation Biology Institute, Corvallis, OR 97333, U.S. Geological Survey, Western Ecological Research Center, Sequoia-Kings Canyon Filed Station, Three Rivers, CA 93271; and Department of Ecology & Evolutionary Biology, University of California, Los Angeles, CA 90095. www.PNAS.org, August 6, 2017. https://www.pnas.org/content/pnas/114/52/13750. full.pdf (Entire paragraph is from this source.)

511 Shellenberger. Ibid. 2019.

512 Shellenberger. Ibid. 2019.

513 Our World in Data, "Life Expectancy," www.OurWorldinData. org, 2019. Page accessed February 15, 2020. https://ourworldindata. org/grapher/life-expectancy?time=1770..2015&country=Africa+ Americas+Asia+Europe+Latin%20America%20and%20the%20 Caribbean+Northern%20America+Oceania+OWID_WRL

514 Roser, Max, Ortiz-Ospina, Esteban, "Literacy," www. OurWorldinData.org, First published in 2013; last revision September 20, 2018. https://ourworldindata.org/literacy

515 Ortiz-Ospina, Esteban, Roser, Max, "Child Labor," www. OurWorldinData.org, 2020. https://ourworldindata.org/child-labor

516 Lomberg, Bjorn, "The Danger of Climate Doomsayers," www.Project-Syndicate.org, August 19, 2019. https://www.project-syndicate.org/ commentary/climate-change-fear-wrong-policies-by-bjorn-lomb org-2019-08

517 Lomberg. Ibid. 2019.

518 Ritchie, Hannah, Roser, Max, "Water Use and Stress," www. OurWorldinData.org, First published in 2015; most recent substantial revision in July 2018. https://ourworldindata.org/water-use-stress

519 Lomberg. Ibid. 2019 and Ritchie & Roser. Ibid. 2018 Water Use and Stress.

520 Lomberg. Ibid. 2019.

521 Ewers, Robert, "Interaction effects between economic development and forest cover determine deforestation rates," www.ScienceDirect. com, Global Environmental Change, Volume 16, Issue 2, May 2006, Pages 161-169. https://www.sciencedirect.com/science/article/abs/pii/ S0959378005000798

522 Rodger, Paul, "Wind turbines generate more cash when switched off and Scottish customers shoulder 650 million (sterling pound)

blame," www.DailyRecord.co.uk, January 17, 2020. https://www.dailyrecord.co.uk/news/scottish-news/wind-turbines-generate-more-cash-21300451

523 Bordoff, Jason, "Big Oil taking up mantra of climate change," www.Reuters.com, February 12, 2020. https://www.reuters.com/article/us-bp-outlook-opinion/big-oil-taking-up-mantle-of-climate-change-idUSKBN20701G

524 Sweeney, Darren, "Virginia lawmakers pass bill phasing out coal generation," www.SpGlobal.com, February 11, 2020. https://www.spglobal.com/marketintelligence/en/news-insights/latest-news-headlines/virginia-lawmakers-pass-bill-phasing-out-coal-generation-57057854

525 Bloomberg Environment, Energy & Environment Report, "Shell, EDP Set Record-Low Price for U.S. Offshore Wind," www.News.BloombergEnvironment.com, February 11, 2020. https://news.bloombergenvironment.com/environment-and-energy/shell-edp-set-record-low-price-for-u-s-offshore-wind-power

526 Flavelle, Christopher, "Global Financial Giants Swear Off Funding an Especially Dirty Fuel," www.NYTimes.com, February 12, 2020. https://www.nytimes.com/2020/02/12/climate/blackrock-oil-sands-alberta-financing.html?ct=t(RSS_EMAIL_CAMPAIGN)

527 Huang, J., Ou, T., Chen, D., Lun, Y., Zhao, Z., "The amplified Arctic warming in the recent decades may have been overestimated by CMIP5 models," www.CO2Science.org, via *Geophysical Research Letters*, pp. 12,345-13,338, 2019. http://www.co2science.org/articles/V23/feb/a6.php

528 Nova, Jo, "Australian grid has major near miss: SA Islanded for two weeks," www.Joannenova.com.au, February 12, 2020. http://joannenova.com.au/2020/02/australian-grid-has-major-near-miss-sa-islanded-for-two-weeks/

529 Jindal, Bobby, "Democrats Want a Prophet, Not a President," www.WSJ.com, (The Wall Street Journal), February 10, 2020. https://www.wsj.com/articles/democrats-want-a-prophet-not-a-president-11581379101

530 Bailey, Ronald, "How Much Will the Green New Deal Cost," www.Reason.com, February 7, 2019. https://reason.com/2019/02/07/green-new-deal-democratic-socialism-by-o/

531 Jindal. Ibid. 2020.

INDEX